T0290980

Numerical Recipes in
Quantum Information
Theory and
Quantum Computing

Numerical Recipes in Quantum Information Theory and Quantum Computing

An Adventure in FORTRAN 90

M. S. Ramkarthik
Payal D. Solanki

CRC Press
Taylor & Francis Group
Boca Raton London New York

CRC Press is an imprint of the
Taylor & Francis Group, an **informa** business

First edition published 2022
by CRC Press
6000 Broken Sound Parkway NW, Suite 300, Boca Raton, FL 33487-2742

and by CRC Press
2 Park Square, Milton Park, Abingdon, Oxon, OX14 4RN

© 2022 M. S. Ramkarthik, Payal D. Solanki

CRC Press is an imprint of Taylor & Francis Group, LLC

The right of M. S. Ramkarthik, Payal D. Solanki to be identified as authors of this work has been asserted by them in accordance with sections 77 and 78 of the Copyright, Designs and Patents Act 1988.

Reasonable efforts have been made to publish reliable data and information, but the author and publisher cannot assume responsibility for the validity of all materials or the consequences of their use. The authors and publishers have attempted to trace the copyright holders of all material reproduced in this publication and apologize to copyright holders if permission to publish in this form has not been obtained. If any copyright material has not been acknowledged please write and let us know so we may rectify in any future reprint.

Except as permitted under U.S. Copyright Law, no part of this book may be reprinted, reproduced, transmitted, or utilized in any form by any electronic, mechanical, or other means, now known or hereafter invented, including photocopying, microfilming, and recording, or in any information storage or retrieval system, without written permission from the publishers.

For permission to photocopy or use material electronically from this work, access www.copyright.com or contact the Copyright Clearance Center, Inc. (CCC), 222 Rosewood Drive, Danvers, MA 01923, 978-750-8400. For works that are not available on CCC please contact mpkbookspermissions@tandf.co.uk

Trademark notice: Product or corporate names may be trademarks or registered trademarks and are used only for identification and explanation without intent to infringe.

Library of Congress Cataloging-in-Publication Data

ISBN: 9780367759285 (hbk)
ISBN: 9780367759315 (pbk)
ISBN: 9781003164678 (ebk)

Typeset in LMR10 font
by KnowledgeWorks Global Ltd.

Access the Support Material: www.routledge.com/

Dedicated to my parents, S. Chitra and R. Seshadri, who developed me from a small piece of biological information to a complex intermixture of ideas, intelligence, and emotions.

M. S. Ramkarthik

Dedicated to my parents Harshaben D. Solanki and Dineshbhai D. Solanki, who encouraged me to excel in every phase of life.

Payal D. Solanki

Contents

Preface

The motivation to write this book was triggered by the fact that the theory of quantum computation and quantum entanglement has a beautiful relation with condensed matter spin systems. In fact, the interplay between both has been so fertile that many topics which was difficult to perceive in these areas have become easier to grasp by this harmonious marriage. During the course of such investigations for several years by the first author, several coding ideas and codes were developed in the FORTRAN programming language. Both authors felt that these codes must be expanded, and optimized with the inclusion of latest computational routines which would be useful in this field. To this end we felt that there was an urgent need of a compendium of such codes written in FORTRAN 90 as a book which can be readily accessed by postgraduate students, practicing physicists, research scholar and experimentalists working in quantum mechanics and quantum information. Such a compendium of various quantum operations borderlining condensed matter systems like spin chains were not in the market when the plan to write such a book was cooking in our minds. So we immediately thought that our hard work in developing these codes must translate into a book with a clear description of various operations in quantum mechanics, quantum information theory, quantum entanglement and spin half condensed matter systems in one dimension. The book was perceived as a encyclopedia of stable, efficient and optimized numerical recipes written in FORTRAN 90 in the aforesaid areas. The reason for choosing FORTRAN as the programming platform for developing the subroutines is due to the fact that it is a highly stable language known for its speed. It is the preferred language for various high performance computers and is used for programs that rank the world's fastest supercomputers. Throughout this book, we have used the open source LAPACK library for performing matrix operations wherever required and we thank and acknowledge the developers of LAPACK.

This book mainly deals with 160 numerical recipes encompassing various operations in the above described areas. We have assumed that the reader is having a good grasp of basic quantum mechanics and quantum information theory and some basic training in programming. Having these in mind, we have spread the book broadly into seven chapters. Each chapter will contain a short theory concerning the recipe, a flowchart describing the flow of the algorithm, an example of calculation and the corresponding output of the recipe. For more deeper knowledge we have provided an extensive bibliography of books and articles to quench the thirst of the interested reader. To avoid confusion and ready usability of the recipes, separate codes for real and complex states have been given as the data types involved are different. The reader has to note that throughout the book we have used the zero based indexing. It is also to be noted that we have used the integer paramter s to represent the total number of qubits in the recipe and the flowcharts. However, in the theoretical description of the recipe we are using N to denote the total number of qubits just as a matter of convention. The flowcharts are designed in a unique way, wherein if there are dependencies in a recipe, we use sub partitions to visualize the input and output variables. To this end, we do a tripartite split of the whole box. For example. The left box contains the input variables written in lower case letters separated by commas like ip1, ip2,... The middle box contains the routine which takes these as the inputs and are written in upper case letters like ROUTINE. Finally the right box contains the output given by the routine in lower case letters separated by commas like op1, op2,... The picture below explains clearly the aforesaid fact,

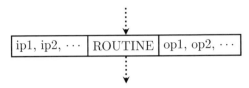

Chapter 1 deals with basics of the FORTRAN 90 programming language. This has been included to give some warm up to the reader who is new to FORTRAN. Every important aspect of FORTRAN used in scientific computation is covered with ideas, syntax, and examples so that it streamlines the thought process of the reader. Several other small but interesting titbits on FORTRAN's features are included. We have also included a list of LAPACK subroutines which we will be using in this book and a thorough discussion of the arguments in these routines. Last but not least, an interesting demonstration of the clock function in FORTRAN is given which will help the user to calculate the computational times for the execution of the full code or a block of the code.

Chapter 2 deals with the nuts and bolts of quantum mechanics, meaning the basic operations which we will be doing with quantum states like normalization, inner and outer products, tensor products, commutators and anti-commutators, inter-conversion between binary and decimal, bit testing, swapping and shifts of qubits. These are the most important and fundamental operations between quantum states.

Chapter 3 deals with numerical linear algebra and matrix operations. Needless to emphasize, the home of these quantum states are the linear vector spaces or alternatively the Hilbert spaces, so this chapter has crucial recipes relating to operations like inverse of matrices, tridiagonalization, QR decomposition, Gram-Schmidt orthogonalization, a wide spectrum of functions and powers of matrices, trace norm, Hilbert-Schmidt norm, singular value decomposition and so on. The routines in this chapter will be used in other chapters, meaning they occur as dependencies in other programs in later chapters. The recipes in this chapter can be not only used in quantum information theory but also be used as standalone recipes for any person working in linear algebra as well.

Chapter 4 deals with the tools of quantum information theory. In this chapter we have given the recipes of all the major quantum gates used in performing various quantum information tasks. The chapter also has some useful recipes on the construction of the N-qubit computational basis, N-qubit GHZ states, N-qubit W states, two-qubit Bell states, N-qubit generalized Werner states. Apart from that, entropy and its quantification by various measures like Shannon entropy, linear and relative entropy are given. It becomes very important in quantum information theory to know the closeness between two quantum states and to this end, we have have given the recipes related to trace distance, fidelity, super fidelity, Bures distance. The chapter ends with two important recipes which are for computing the expectation values of observables and the quantum mechanical single qubit measurements.

Chapter 5 deals with quantum entanglement and its quantification only for qubit systems. The reason why we have devoted a complete chapter on quantum entanglement measures and detection of entanglement is because there are a wide variety of entanglement measures for pure and mixed states which are used very often. Before going into the chapter, two main and indispensable operations namely the partial tracing and the partial transpose are explained and the recipes are given in detail. The interesting feature of these routines is the fact that, these operations can be done between any random set of qubits not necessarily contiguous qubits. Then follows the recipes to calculate the entanglement measures like concurrence and its variants, von Neumann entropy, block entropy, Renyi entropy, Q measure, negativity and log negativity, entanglement spectrum and residual entanglement (three tangle). Last but not least, certain entanglement detection procedures like PPT (the Peres-Horodecki criteria) and the reduction criteria are also explained via examples.

Chapter 6 deals with the construction of the Hamiltonians of one-dimensional spin half models in condensed matter physics. Hamiltonian of spins interacting with an external magnetic field (both homogeneous and inhomogeneous) and separate Hamiltonians for the direct exchange spin-spin interactions in X, Y and Z directions, Heisenberg Hamiltonian and the Dzyaloshinskii-Moriya interaction are dealt with in detail. We emphasize here that the Hamiltonians which we have given are not only for the nearest-neighbour interaction but it is for the rth nearest-neighbour interaction. Another unique feature of these recipes is that, it can be used to construct the Hamiltonians with both isotropic and random interaction strengths. The aforesaid features increases the scope and usability of the recipes. Other important variants such as the Ising model, transverse Ising model, Majumdar-Ghosh model and a plethora of other one-dimensional spin chain models can be easily constructed using our recipes given.

The last chapter deals with the construction of arrays of Gaussian or uniformly distributed random numbers. Using them, we construct important quantum mechanical objects like random quantum vectors and states, random density matrices, and random matrices like Hermitian, unitary, Ginibre and Wishart. This chapter was perceived because random quantum states play a very important role not only in quantum information theory but also in areas such as nuclear physics.

This book would not have seen the light of day without the help of Ms. Carolina Antunes of CRC Press Taylor & Francis who at every stage of the work was very supportive to us. I would also like to thank Ms. Lara Speiker of the Editorial department for being extremely patient in answering all our questions related to the typesetting and the general layout of the book. I would like to thank CRC Press, Taylor & Francis for giving us this opportunity to write this book and make the recipes available to all aspiring researchers in this area. Last but not least, the recipes given in this book were verified several times with different physical systems and to the best of our knowledge gives accurate outputs, in spite of several rounds of careful editing and testing, there may always be errors for which the authors will feel grateful if the readers point out the same.

M. S. Ramkarthik
Payal D. Solanki

List of Figures

Acknowledgment

I would like to acknowledge several people, first and foremost among them are my parents who supported me in all possible ways to become a physicist and they are my first teachers. I would like to thank my wife for being very concerned about my growth and who has been a pillar of support in my very hard times. My sincere thanks goes to my high school physics teacher Mrs. A. K. Mythili who not only taught me physics in its pristine form but also taught me scientific honesty, those seeds sown at the young age developed as a tree now bearing fruits upto whatever capacity it can. I thank my beloved professors Dr. S. S. Vasan, Dr. M. S. Sriram, Dr. M. S. Raghunathan, Dr. M. V. Satyanarayana, Dr. K. Sethupathi for sharing their knowledge to me in the most friendly way it can be making learning a wonderful experience. It is my pleasure to thank Prof. Pramod Padole, Director VNIT and Prof. Dilip Peshwe for their support. It is my duty to thank all my students, be it undergraduate, postgraduate or PhD scholars whom I have been teaching at my institute for the past several years, it was their questions and interactions which further polished my abilities. Finally I thank the almighty God for being with me as a guiding light in the form described as 'Virupaksha' meaning 'consciousness that has eyes but no form'.

M. S. Ramkarthik

First and foremost, I would like to acknowledge my parents who have been continuous source of inspiration and strength to me. I would also like to express my gratitude toward my high school mathematics teacher Mr. Paresh Lakhani, and my graduate professors Prof. Viresh Thakkar and Dr. Pruthul Desai, under whose guidance I developed a keen interest in the fields of mathematics and quantum mechanics, without their blessings it would have been impossible for me to further excel in this field. I would like to especially thank my Ph.D. thesis supervisor Dr. M. S. Ramkarthik for giving me this opportunity to be a co author. His dedication and honesty toward work has inspired me to become a good researcher. This acknowledgment cannot be completed without showing my appreciation for my friend Pranay Barkataki, whose constructive criticism on the numerical recipes of the book provided me invaluable insights. Finally I would like to thank my beloved friend Aprajita Joshi, with whom I shared the excitement of learning and the bond of friendship.

Payal D. Solanki

1

Introduction to FORTRAN 90

For those who are new to the FORTRAN [1–3] programming language this chapter will give a detailed explanation of the FORTRAN 90 language, its syntax and structure. In this chapter we give a brief introduction to the FORTRAN 90 programming language. This we feel is important because it can provide a ready reference for those who use this book. FORTRAN is an open source distribution wherein anyone is free to download and use it. FORTRAN was developed by John Backus and his team at IBM [4] in 1957 specifically designed for scientific programming. In fact, the name FORTRAN stands for FORmula TRANslation and the FORTRAN code is more mathematically readable. From then on, the language has evolved exponentially with various features being updated in every release of it. The modern version of FORTRAN has many new features in it. However, we will be adhering to FORTRAN 90 throughout this book.

In the further sections we will give a complete view of FORTRAN 90. The program is written in any of the infinite text editors available in the web market and this program file should have file extension as '.f90' for the compiler to recognize it. All the programs in FORTRAN start with the keyword PROGRAM and it ends with END PROGRAM. FORTRAN uses 'implicit real(a - h, o - z)' reserves i to n as an integer by default and the complementary alphabets as real. However, if we use 'implicit none', it allows us to define a variable of our choice. The authors feel that implicit none is better as we will have control over our variables declared and it will facilitate easy debugging, however this is purely the choice of the programmer. FORTRAN codes are case insensitive except for string literals. Also note that FORTRAN handles single and double precision with equal ease. The single precision has 8 significant digits and double precision has 16 significant digits after the decimal place. The gross frame of a FORTRAN program is shown below.

```
program name
implicit none
.......
body of the program
.......
end program
```

We will be using the gfortran compiler here, throughout the book.

1.1   Printing on screen

No we will see how to print some character or a number on screen, i.e. on the terminal which we are working on. Syntax for printing 'hello' on the terminal is given below,

```
print*, 'hello'
```

1.2   Data types

1.2.1   Integer

Integer data type can only hold integer numbers and we can define the integer data type as given below:

```
integer::a
```

where a is an integer. For different byte size of integer we should define integer type as,

```
integer::a
integer*2::b
integer*4::c
integer*8::d
```

Type of the default integer is 'integer*4'.

Data type	Number of bits	Range
integer*1	8	-128 to $+127$
integer*2	16	$-32,768$ to $+32,767$
integer*4	32	$-2,147,483,648$ to $+2,147,483,647$
integer*8	64	-9.22×10^{18} to $+9.22 \times 10^{18}$

1.2.2   Real

Real data types store floating point numbers (real numbers). There are two kinds of floating point numbers **real (real*4)** and **double precision (real*8)**.

```
real::a
real*8::b
double precision::c
```

Where single precision data type contains 8 digits after decimal point, whereas double precision contains 16 digits after decimal point. Whenever we want to convert any integer value or a single precision real value to a double precision value during assignment in the main program, we can write the number followed by '**d0**'. This can more easily be understood by an example given below:

```
program test
real*4::a
real*8::b,c
a=52.3
b=52.3
c=52.3d0
print*,a,b,c
end program
```

The above code will print the value of a, b and c, respectively as,

```
52.2999992 52.299999237060547 52.299999999999997
```

Thus we see that when we use 'd0', we get double precision accuracy which is the closest value to the input value 52.3.

1.2.3 Complex

Any complex number has two parts, real and imaginary and it has two types, complex for single precision and complex*16 for double precision.

```
complex::a
complex*16::b
```

Syntax for assigning the complex number $1 + 3i$ to any variable is given below:

```
program test
complex*8::a
complex*16::b
a=cmplx(1.d0,3.d0)
b=dcmplx(1.d0,3.d0)
print*,a
print*,b
end program
```

where cmplx and dcmplx is for single and double precision complex numbers, respectively. The output of the above code in single and double precision, respectively are,

```
(1.00000000,3.00000000)
(1.0000000000000000,3.0000000000000000)
```

1.2.4 Character

Character data type stores string and characters which are words or phrases as given in the example below, where **len** specifier is for the length of the strings.

```
program test
character(len=7)::s
s='physics'
print*,s
end program
```

Which will print the output on the terminal as follows,

```
physics
```

1.2.5 Logical

There are only two logical values **.True.** and **.False.** corresponding to the standard Boolean elements 1 and 0, respectively. An example is indicated as below:

```
program test
logical:: i
i=.false.
print*,i
end program
```

Which will print 'F' on terminal.

1.3   Constants

Constants refer to the fixed values that the program cannot alter during its execution. All the constants in a program can be declared using the parameter command in FORTRAN which is very useful. This parameter can be of any data type and it can be an array too of any dimension, however, we explicitly give an example to define a integer or a real parameter as follows.

```
program test
integer,parameter::k=2
real,parameter::g=9.8
end program
```

Any number of parameters can be declared in the preamble of the program. The advantage of using the parameter function is that, quickly we can change the inputs in program globally without altering the program much.

1.4   Operators

Operators are objects which perform certain mathematical operations to give another output. Here A and B are called operands. There are three types of operators which can be defined in FORTRAN which are the arithmetic, relational and logical operators as we will see further.

1.4.1   Arithmetic operators

Operator	Description
+	Adds two operands, i.e. $A + B$
-	Subtracts two operands, i.e. $A - B$
*	Multiply two operands, i.e. $A * B$
/	Performs division, i.e. A/B
**	Raises one operand to the power of the other, i.e. $A ** B = A^B$

1.4.2 Relational operators

Operator	Description
== or .eq.	It checks whether two operands have the same value or not.
/= or .ne.	It checks whether two operands have equal values or not.
> or .gt.	It checks whether the left operand is greater than the right operand or not.
< or .lt.	It checks whether the left operand is less than the right operand or not.
>= or .ge.	It checks whether the left operand is greater than or equal to the right operand or not.
<= or .le.	It checks whether the left operand is less than or equal to the right operand or not.

1.4.3 Logical operators

Operator	Description
.and.	It is called the logical AND operator. If both the operands are non-zero, then the condition becomes true.
.or.	It is called the logical OR operator. If any one of the two operands are non-zero, then the condition becomes true.
.not.	It is called the logical NOT operator. Used to reverse the logical state of its operand. If a condition is true then the logical NOT operator will make false and vice versa.
.eqv.	It is called the logical EQUIVALENT Operator. Used to check the equivalence between two logical values.
.neqv.	It is called the logical NON-EQUIVALENT operator. Used to check the non-equivalence between two logical values.

1.5 Decisions

Decisions are very important elements in any programming language. Conditional statements helps us in making those decisions. Basic syntax for conditional statements are given below,

1.5.1   If... then... end if

```
if (logical expression) then
   statements
end if
```

An example of the above syntax is given below:

```
program test
implicit none
integer::a
a=3
if (a<5) then
    a=a+2
end if
print*,a
end program
```

Which will print the output as,

```
5
```

1.5.2   If...then...else

```
if (logical expression) then
   statements
else
   Other statements
end if
```

An example of the above syntax is given below:

```
program test
implicit none
integer::a
a=8
if (a<5) then
    a=a+2
else
    a=a-2
end if
print*,a
end program
```

Which will print the output as,

```
6
```

1.6   Loops

Loops are useful for repeated execution of similar operations. The importance of looping is so much that, without looping elements any programming language will be 99% incomplete. There are two types of loops and syntax of both of them are given below:

1.6.1   The do loop

```
do var=start,stop,stepsize
statements
end do
```

Where var is any loop variable. In the example given below, 'i' is the loop variable.

```
program test
implicit none
integer::a,i
a=3
do i=0,5,1
    a=a+i
end do
print*,a
end program
```

Which will print the output as,

```
18
```

It is not necessary that the loop must be only incremental, it can be decremental as well. This is illustrated below, we start the counter with 5 and decreasing in steps of 1 which is indicated by −1.

```
program test
implicit none
integer::a,i
a=3
do i=5,1,-1
    a=a+i
end do
print*,a
end program
```

Which will print the output as,

```
18
```

When we use the print statement after the 'end do' statement, last step of the loop execution will only be printed. However in certain cases we require to print every step of the loop for which we have to put the 'print' command before the 'end do' statement. In the example below, the above point is highlighted.

```
program test
implicit none
integer::i
do i=1,3,1
    print*,i*i
end do
end program
```

Which will print the output as,

```
1
4
9
```

1.6.2   The do while loop

```
do while (logical expression)
    statements
end do
```

The example of the above syntax is given below:

```
program test
implicit none
integer::a,n
a=3
n=0
do while (n<=5)
    a=a+n
    n=n+1
end do
print*,a
end program
```

Which will print output as,

```
18
```

Note that the loop variable always should be an integer in FORTRAN 90. While writing any program, real stpe size loops are very important specially for solving any scientific problem. The method of constructing the real step size loop is shown below with an example. This is so because, in FORTRAN 90, the loops should have integer step size only. So a small tweak will suffice.

```
program test
implicit none
integer::i
real*8::j
do i=0,10,1
    j=i/10.0d0
    print*,j
end do
```

```
end program
```

In the above example we have used a dummy variable j for the real step size. It will print j from 0 to 1 in steps of 0.1. However, the program given below will print j from 0 to 1 in step of 0.01.

```
program test
implicit none
integer::i
real*8::j
do i=0,100,1
    j=i/100.0d0
    print*,j
end do
end program
```

1.7   EXIT, STOP and CYCLE

1.7.1   EXIT

The EXIT statement terminates the loop depending on the conditions. The program immediately comes out from the loop and performs the next statements succeeding the loop.

```
program test
implicit none
integer::i,a
a=3
do i=0,10,1
    if (i>5) then
        exit
    end if
    a=a+i
end do
print*,a
end program
```

Which prints output as,

```
18
```

However without the EXIT statement it will produce the output as 58.

1.7.2   STOP

STOP statement aborts the execution and immediately finishes the program.

```
program test
implicit none
integer::i,a
a=3
```

```
do i=0,10,1
    if (i>5) then
        STOP
    end if
    a=a+i
end do
print*,a
end program
```

Which will not print anything because we have given the print statement after the stop statement. However without the STOP statement the output will be 58.

1.7.3   CYCLE

The cycle statement causes the loop to skip the current iteration when it satisfies the given condition.

```
program test
implicit none
integer::i
do i=0,10,1
    if (i>2 .and. i<8) then
        CYCLE
    end if
    print*,i
end do
end program
```

Which will print i as,

0
1
2
8
9
10

The values 3, 4, 5, 6, 7 are missing as the case should be dictated by the condition $i > 2$ and $i < 8$. The EXIT, STOP and CYCLE features can be useful while debugging the program.

1.8   Arrays

An array is a collection of items stored in contiguous memory locations. They can be imagined like matrices in mathematics, where numbers are stored in a one or two-dimensional arrangement. Arrays mostly are used to store the collection of data which are of the same type. The lowest address corresponds to the first element and the highest address corresponds to the last element. Arrays can be one-dimensional (like vectors), two-dimensional (like matrices). Note that FORTRAN allows us to create up to seven-dimensional arrays, however most frequently we use upto two dimensions or in rare cases higher dimensions depending on the problem and the method of solution under consideration.

1.8.1 Declaration

There are two types of indexing in arrays which are,

1. *Zero-based indexing*: The index of first element starts with zero, an example of such a real array is given below:

```
real,dimension(0:5)::A
real,dimension(0:5,0:5)::C
```

   Where 0 is the lower bound and 5 is the upper bound for the array.

2. *One-based indexing*: The index of the first element starts with one, the example of such an integer array is given below:

```
integer,dimension(6)::B
integer,dimension(6,6)::D
```

   Here the lower bound and upper bound of the arrays are 1 and 6, respectively.

A noteworthy point here is, in this book we are following the zero-based indexing only

1.8.2 Assignment

Usually we can directly assign the values to an array or we can use loops for the assignment. An example of an assignment is given below. For example, let us take a matrix A of the dimension 3×3 as,

$$A = \begin{pmatrix} 1 & 2 & 3 \\ 1 & 2 & 3 \\ 1 & 2 & 3 \end{pmatrix} \tag{1.1}$$

We can assign this matrix using a loop as shown below:

```
program test
implicit none
integer::i,j,A(3,3)
do i=1,3,1
    do j=1,3,1
        A(i,j)=j
    end do
end do
end program
```

Second method for similar assignment is given below, where we assign the individual elements separately.

```
program test
implicit none
integer::A(0:2,0:2)
A(0,:)=(/1,2,3/)
A(1,:)=(/1,2,3/)
A(2,:)=(/1,2,3/)
end program
```

Initialization of array with all entries to be zero is very important when we have many zero elements in the matrix, and only a few elements have to be populated, basically this procedure is good when we deal with sparse matrices.

```fortran
program test
implicit none
integer::i,A(3,3)
A=0
A(1,1)=1
A(2,1)=3
A(2,2)=2
A(3,3)=3
do i=1,3
    print*,A(i,:)
end do
end program
```

Which will print output as,

1	0	0
3	2	0
0	0	3

It has to be noted here that in the above program, The statement $A(i,:)$ will print the output in the matrix form as the required two-dimensional array. Sometimes such a visual output in matrix form is required for verification and debugging purposes. Apart from this, it is mathematically pleasing to look at.

1.9   Dynamic arrays

Dynamic arrays are special arrays which continuously change their dimensions during the execution of the program. The size of the dynamic array is not known prior to compilation. Dynamic array can be declared using the attribute called **allocatable**, an example of a one-dimensional integer dynamic array A and a real double precesion two-dimensional dynamic array B is shown as,

```fortran
integer,dimension(:),allocatable::A
real*8,dimension(:,:),allocatable::B
```

For allocating the memory to this dynamic array we use the function called **allocate**. Let us allocate a dimension p to this array A and $m \times m$ dimension to array B as,

```fortran
allocate(A(p),B(m,m))
```

De allocation of the array can be done using the function called **deallocate** as,

```fortran
deallocate(A,B)
```

All the calculations corresponding to dynamic arrays are done between two functions **allocate** and **deallocate**, as once the array gets deallocated, all the entries and the whole array will be erased from the memory. Note that, the example given above is for an one-dimensional array A (with one based indexing), however this can be used for any dimensions and also with any indexing. Since allocatable uses dynamic memory allocation, it should be used with extreme care. We have given an example of using a one-dimensional allocatable array as follows.

```
program test
implicit none
integer,parameter::p=3
integer,dimension(:),allocatable::a
integer::i
allocate (a(p))
do i=1,p
   a(i)=i
end do
print*,a
deallocate(a)
end program
```

Which prints output as,

1	2	3

1.10 Reading and writing

1.10.1 Writing in the file

Syntax for writing some data into a file is given below:

```
write(111,*) 'hello'
```

Where 'hello' will be written in the file called **fort.111**. Instead of 111 we can use other number also. Note that write(6,*) will print the output on screen (on terminal). The syntax of writing three variables a, b and c is given as follows,

```
write(111,*) a,b,c
```

1.10.2 Reading from file

Let us take an example, of the file named as fort.111 and five real numbers are written in that file. We want to read those numbers in some other program as its input. The syntax for storing these five numbers in an one-dimensional array 'a' as follows.

```
do i=1,5,1
   read(111,*)a(i)
end do
```

A similar procedure can be followed for two-dimensional array also.

1.11   Intrinsic functions

FORTRAN has predefined functions in its compiler for ready use, these are called intrinsic functions and they can be directly called in the program. We have tabulated some of the frequently used intrinsic functions as given below.

Operator	Description
ABS(A)	It returns the absolute value of A.
AIMAG(Z)	It returns the imaginary part of a complex number Z.
REAL(Z)	It returns the real part of a complex number Z.
CONJG(Z)	It returns the complex conjugate of any complex number Z.
MOD(A, P)	It returns the remainder of A on division by P.
COS(X)	It returns the cosine of an argument in radians.
SIN(X)	It returns the sine of an argument in radians.
TAN(X)	It returns the tangent of an argument in radians.
ACOS(X)	It returns the inverse cosine in the range $(0, \pi)$, in radians.
ASIN(X)	It returns the inverse sine in the range $(-\pi/2, \pi/2)$, in radians.
ATAN(X)	It returns the inverse tangent in the range $(-\pi/2, \pi/2)$, in radians.
EXP(X)	It returns the exponential of X.
LOG(X)	It returns the natural logarithm of X (that is to the base e).
SQRT(X)	It returns the square root of X.
SINH(X)	It returns the hyperbolic sine of the argument in radians.
COSH(X)	It returns the hyperbolic cosine of the argument in radians.
TANH(X)	It returns the hyperbolic tangent of the argument in radians.
LOG10(X)	It returns the common logarithmic to the (base 10) value of X.
INT(X)	It returns the integer part of the integer, real or complex variable X.
DFLOAT(X)	It returns the double precision value of integer X.

However while using double precision, all the intrinsic function will carry a 'D' before it. For example DSIN(X), etc.

1.12 Subroutines

If the FORTRAN codes are big and spanning several lines, debugging will be very difficult. In these contexts, if we break the program into several parts, the life of the programmer will be relatively easier as the program will look more uncluttered. This work is done by defining the subroutines in the program. Many times it happens that, some part of the program involving a particular operation repeats several times, then, these operations are common to the entire program and it can be defined as a subroutine and called in the main program wherever required. Subroutines typically are collected after the END PROGRAM statement or it can be called from other FORTRAN files too. The recipes we have given in this book are in the form of the subroutine. The syntax for the subroutine is given below:

```
subroutine name(arg1, arg2, ....)
    declarations, including those for the arguments
    executable statements
end subroutine
```

arg1, arg2,... will be the input and output variables. Note that subroutine can have any number of arguments. We have illustrated the idea of a subroutine by using it to find the factorial of an integer as given below:

```
subroutine factorial(n,factn)
implicit none
integer::factn,n,i
factn=1
do i = 1, n
    factn=factn*i
end do
end subroutine
```

In the above subroutine, n is the input parameter and factn is the output parameter. We need to call subroutines into the main program, syntax corresponding to calling is given below.

```
program example
integer::factn
call factorial(4,factn)
print*,factn
end program
```

Which will print the output as 24 which is 4!.

1.13 Some tit bits in FORTRAN for effective programming

Here we give some useful list of procedures which can be handy for a scientific programmer. All these features are included in a program called 'titbit' as below. Here we define the most useful constants like e, π and $i = \sqrt{-1}$. If we define it this way, we have given a free hand to the machine to take up any precision of these constants.

```
program titbit
implicit none
real*8::e,pi
complex*16::i
e=exp(1.0d0)
pi=4.0d0*atan(1.0d0)
i=dcmplx(0,1)
print*,e,pi
print*,i
end program
```

Which will print the output as,

```
2.7182818284590451 3.1415926535897931
     (0.0000000000000000,1.0000000000000000)
```

1.14 LAPACK

In this book, calculations involving matrix operations are done by the open source library called LAPACK (Linear Algebra PACKage) and BLAS (Basic Linear Algebra Subprograms). These two libraries have many subroutines written in FORTRAN 90 and provides routines for solving systems of simultaneous linear equations, least-squares solutions of linear systems of equations, matrix eigenvalue problems, singular value problems, etc. The associated matrix factorizations like (LU, Cholesky, QR, SVD, Schur, generalized Schur) are also provided. These routines cater for real and complex matrices which are both dense and banded, in both single and double precision. Each subroutine has a characteristic name associated with it. In this section we give a detailed analysis of how to use these subroutines, of course these are now black boxes as we are just interested in using them without going into the nitty-gritties of the actual algorithm. The convention adopted here is - [in] means that the parameter need to be fed into the routine by the user and [out] means the parameter will be given by the routine to the user, however [ws] stands for workspace. For detailed understanding, the user can refer the LAPACK manual [5] or the netlib website. Here we remark that only those subroutines which we have used in this book are described in the following sections to suit our needs.

1.14.1 DSYEV

Syntax

```
call DSYEV( JOBZ, UPLO, N, A, LDA, W, WORK, LWORK, INFO )
```

Purpose

DSYEV computes all the eigenvalues and optionally the eigenvectors of a real symmetric matrix A.

Arguments

Argument	In/Out	Type	Description
JOBZ	[in]	character*1	= 'V': Computes eigenvalues and eigenvectors; = 'N': Computes only eigenvalues.
UPLO	[in]	character*1	= 'U': Upper triangle of A is stored; = 'L': lower triangle of A is stored.
N	[in]	integer	It is the dimension of the matrix we are dealing with.
A	[in/out]	real*8	A is the N × N input matrix. If JOBZ = 'V', on exit, the columns of A will be orthonormal eigenvectors. If JOBZ = 'N', then on exit, the lower triangle (if UPLO = 'L') or the upper triangle (if UPLO = 'U') of A, including the diagonal, is destroyed.
LDA	[in]	integer	LDA is the leading dimension of array A which is N.
W	[out]	real*8	W is an array of dimension N and it stores eigenvalues in ascending order.
LWORK	[in]	integer	LWORK should be greater than or equal to $3N-1$.
WORK	[ws/out]	real*8	WORK is an array of dimension equal to LWORK. On exit, if INFO = 0, WORK(0) returns the optimal LWORK.
INFO	[out]	integer	This parameter is for crosschecking the code. If INFO = 0, successful exit.

1.14.2 ZHEEV

Syntax

```
call ZHEEV( JOBZ, UPLO, N, A, LDA, W, WORK, LWORK, RWORK,INFO )
```

Purpose

ZHEEV computes all the eigenvalues and optionally the eigenvectors of a complex Hermitian matrix A.

Arguments

Argument	In/Out	Type	Description
JOBZ	[in]	character*1	= 'V': Computes eigenvalues and eigenvectors; = 'N': Computes only eigenvalues.
UPLO	[in]	character*1	= 'U' : Upper triangle of A is stored; = 'L': Lower triangle of A is stored.
N	[in]	integer	It is the dimension of the matrix we are dealing with.
A	[in/out]	complex*16	A is the N × N input matrix. If JOBZ = 'V', on exit, the columns of A will be orthonormal eigenvectors. If JOBZ = 'N' , on exit, the lower triangle (if UPLO = 'L') or the upper triangle (if UPLO = 'U') of A, including the diagonal, is destroyed.
LDA	[in]	integer	LDA is the leading dimension of the array A which is N.
W	[out]	real*8	W is an array of dimension N and it stores eigenvalues in ascending order.
LWORK	[in]	integer	LWORK should be greater than or equal to $2N-1$.
WORK	[ws/out]	complex*16	WORK is an array of dimension equal to LWORK. On exit, if INFO = 0, WORK(0) returns the optimal LWORK.
RWORK	[ws]	real*8	RWORK is an array of dimension $3N-2$.
INFO	[out]	integer	This parameter is for crosschecking the code. If INFO = 0, successful exit.

1.14.3   DGEEV

Syntax

```
call DGEEV( JOBVL, JOBVR, N, A, LDA, WR, WI, VL, LDVL, VR, LDVR, WORK,
    LWORK, INFO )
```

Purpose

DGEEV computes all the eigenvalues and optionally the left or/and right eigenvectors of a real non-symmetric matrix A.

Arguments

Argument	In/Out	Type	Description
JOBVL	[in]	character*1	= 'N': left eigenvectors of A are not computed; = 'V': left eigenvectors of A are computed.
JOBVR	[in]	character*1	= 'N': right eigenvectors of A are not computed; = 'V': right eigenvectors of A are computed.
N	[in]	integer	It is the dimension of the matrix we are dealing with.
A	[in/out]	real*8	A is the N × N input matrix and on exit, A will be overwritten.
LDA	[in]	integer	LDA is the leading dimension of the array A which is N.
WR	[out]	real*8	WR is an array of dimension N and it stores the real part of the eigenvalue.
WI	[out]	real*8	WI is an array of dimension N and it stores the imaginary part of the eigenvalue.
VL	[out]	real*8	VL is an array of the dimension LDVL × N. • If JOBVL = 'V', the left eigenvectors are stored one after another in the columns of VL, in the same order as their eigenvalues. • If the jth eigenvalue is real, then the jth column of VL corresponds to the jth eigenvector. If the jth and (j+1)th eigenvalues form a complex conjugate pair, [VL(:,j) + i*VL(:,j+1)] corresponds to the jth eigenvector and [VL(:,j) − i*VL(:,j+1)] corresponds to the (j+1)th eigenvector. • If JOBVL = 'N', VL is not referenced.
LDVL	[in]	integer	LDVL is the leading dimension of array. LDVL ≥ 1; if JOBVL = 'V', LDVL ≥ N.
VR	[out]	real*8	VR is an array of dimension LDVR × N. • If JOBVR = 'V', the right eigenvectors are stored one after another in the columns of VR, in the same order as their eigenvalues.

Argument	In/Out	Type	Description
			• If the jth eigenvalue is real, then the jth column of VR corresponds to the jth eigenvector. If the jth and (j+1)th eigenvalues form a complex conjugate pair, [VR(:,j) + i*VR(:,j+1)] corresponds to the jth eigenvector and [VR(:,j) − i*VR(:,j+1)] corresponds to the (j+1)th eigenvector. • If JOBVR = 'N', VR is not referenced.
LDVR	[in]	integer	The leading dimension of the array VR. LDVR $\geq$ 1; if JOBVR = 'V', LDVR $\geq$ N.
LWORK	[in]	integer	LWORK is the dimension of array WORK. LWORK $\geq$ 3N, and if JOBVL = 'V' or JOBVR = 'V', LWORK $\geq$ 4N.
WORK	[ws/out]	real*8	WORK is the array of dimension equal to LWORK. On exit, if INFO = 0, WORK(0) returns the optimal LWORK.
INFO	[out]	integer	This parameter is for crosschecking the code. If INFO = 0, successful exit.

1.14.4 ZGEEV

Syntax

```
call ZGEEV( JOBVL, JOBVR, N, A, LDA, W, VL, LDVL, VR, LDVR, WORK, LWORK,
    RWORK, INFO )
```

Purpose

ZGEEV computes all the eigenvalues and optionally left or/and right eigenvectors of a complex non-symmetric matrix A.

Arguments

Argument	In/Out	Type	Description
JOBVL	[in]	character*1	= 'N': left eigenvectors of A are not computed; = 'V': left eigenvectors of A are computed.
JOBVR	[in]	character*1	= 'N': right eigenvectors of A are not computed; = 'V': right eigenvectors of A are computed.

Argument	In/Out	Type	Description
N	[in]	integer	It is the dimension of the matrix we are dealing with.
A	[in/out]	complex*16	A is N × N input matrix and on exit on exit *A* will be overwritten.
LDA	[in]	integer	LDA is the leading dimension of array A which is N.
W	[out]	complex*16	W is the array of dimension N and it stores eigenvalues.
VL	[out]	complex*16	VL is array of the dimension LDVL × N. If JOBVL = 'V', the left eigenvectors are stored one after another in the columns of VL, in the same order as their eigenvalues. If JOBVL = 'N', VL is not referenced.
LDVL	[in]	integer	The leading dimension of the array VL. LDVL ≥ 1; if JOBVL = 'V', LDVL ≥ N.
VR	[out]	complex*16	VR is array of the dimension LDVR × N. If JOBVR = 'V', the right eigenvectors are stored one after another in the columns of VR, in the same order as their eigenvalues. If JOBVR = 'N', VR is not referenced.
LDVR	[in]	integer	LDVR is the leading dimension of the array VR. LDVR ≥ 1; if JOBVR = 'V', LDVR ≥ N.
LWORK	[in]	integer	LWORK should be greater than or equal to 2N.
WORK	[ws/out]	complex*16	WORK is an array of dimension equal to LWORK. On exit, if INFO = 0, WORK(0) returns the optimal LWORK.
RWORK	[ws]	real*8	RWORK is an array of the dimension 2N.
INFO	[out]	integer	This parameter is for crosschecking the code. If INFO = 0, successful exit.

1.14.5 DGEQRF

Syntax

```
call DGEQRF( M, N, A, LDA, TAU, WORK, LWORK, INFO )
```

Purpose

DGEQRF computes the QR decomposition of the $M \times N$ real matrix A.

Arguments

Argument	In/Out	Type	Description
M	[in]	integer	It is the number of rows in A.
N	[in]	integer	It is the number of columns in A.
A	[in/out]	real*8	A is the M × N input matrix. On exit, the elements on and above the diagonal of the array contain the min(M,N) × N upper trapezoidal matrix R (R is upper triangular if M ≥ N); the elements below the diagonal, with the array TAU, represents the orthogonal matrix Q as a product of min(M,N) elementary reflectors.
LDA	[in]	integer	LDA is the leading dimension of the array A which is M.
TAU	[out]	real*8	TAU is an array of dimension min(M,N) and it stores scalars which will be helpful in finding reflectors.
LWORK	[in]	integer	LWORK should be greater than or equal to N.
WORK	[ws/out]	real*8	WORK is an array of dimension equal to LWORK. On exit, if INFO = 0, WORK(0) returns the optimal LWORK.
INFO	[out]	integer	This parameter is for crosschecking the code. If INFO = 0, successful exit.

1.14.6   DORGQR

Syntax

```
call ( M, N, K, A, LDA, TAU, WORK, LWORK, INFO )
```

Purpose

DORGQR generates an $M \times N$ real matrix Q with orthonormal columns. DORGQR and DGEQRF collectively being used for QR decomposition and the output of DGEQRF will be fed into DORGQR which will result finally in the orthonormal matrix Q.

Arguments

Argument	In/Out	Type	Description
M	[in]	integer	It is the number of rows in A.

Argument	In/Out	Type	Description
N	[in]	integer	It is the number of columns in A.
K	[in]	integer	K is the number of reflectors which is N.
A	[in/out]	real*8	Feed the output of DGEQRF as A which is the M × N input matrix and on exit it will be an orthonormal matrix Q.
LDA	[in]	integer	LDA is a first dimension of array A which is M.
TAU	[in]	real*8	TAU is an array of dimension min(M,N), it must contain scalar factors which is returned by DGEQRF.
LWORK	[in]	integer	LWORK should be greater than or equal to N.
WORK	[ws/out]	real*8	WORK is an array of dimension equal to LWORK. On exit, if INFO = 0, WORK(0) returns the optimal LWORK.
INFO	[out]	integer	This parameter is for crosschecking the code. If INFO = 0, successful exit.

1.14.7 ZGEQRF

Syntax

```
call ZGEQRF( M, N, A, LDA, TAU, WORK, LWORK, INFO )
```

Purpose

ZGEQRF computes the QR decomposition of the $M \times N$ complex matrix A.

Arguments

Argument	In/Out	Type	Description
M	[in]	integer	It is the number of rows in A.
N	[in]	integer	It is the number of columns in A.
A	[in/out]	complex*16	A is M × N input matrix. On exit, the elements on and above the diagonal of the array contain the min(M,N) × N upper trapezoidal matrix R (R is upper triangular if M ≥ N); the elements below the diagonal, with the array TAU, represent the unitary matrix Q as a product of min(M,N) elementary reflectors.

Argument	In/Out	Type	Description
LDA	[in]	integer	LDA is the leading dimension of array A which is M.
TAU	[out]	complex*16	TAU is an array of dimension min(M,N) and it stores scalar which will be helpful in finding reflectors.
LWORK	[in]	integer	LWORK should be greater than or equal to N.
WORK	[ws/out]	complex*16	WORK is an array of dimension equal to LWORK. On exit, if INFO = 0, WORK(0) returns the optimal LWORK.
INFO	[out]	integer	This parameter is for crosschecking the code. If INFO = 0, successful exit.

1.14.8   ZUNGQR

Syntax

```
call ( M, N, K, A, LDA, TAU, WORK, LWORK, INFO )
```

Purpose

ZUNGQR generates an $M \times N$ complex matrix Q with orthonormal columns. ZUNGQR and ZGEQRF collectively being used for QR decomposition and the output of ZGEQRF will be fed into ZUNGQR which will finally result in the unitary matrix Q.

Arguments

Argument	In/Out	Type	Description
M	[in]	integer	It is the number of rows in A.
N	[in]	integer	It is the number of columns in A.
K	[in]	integer	K is the number of reflectors which is N.
A	[in/out]	complex*16	Feed the output of ZGEQRF as A which is M $\times$ N input matrix and on exit it will be unitary matrix Q.
LDA	[in]	integer	LDA is the first dimension of array A which is M.
TAU	[in]	complex*16	TAU is an array of dimension min(M,N), it must contain scalar factors which will be returned by ZGEQRF.

Argument	In/Out	Type	Description
LWORK	[in]	integer	LWORK should be greater than or equal to N.
WORK	[ws/out]	complex*16	WORK is an array of dimension equal to LWORK. On exit, if INFO = 0, WORK(0) returns the optimal LWORK.
INFO	[out]	integer	This parameter is for crosschecking the code. If INFO = 0, successful exit.

1.14.9   DSYTRD

Syntax

```
call DSYTRD( UPLO, N, A, LDA, D, E, TAU, WORK, LWORK, INFO )
```

Purpose

DSYTRD reduces a real symmetric matrix A into a tridiagonal matrix T by an orthogonal similarity transformation.

$$O \, A \, O^T = T \qquad (1.2)$$

Arguments

Argument	In/Out	Type	Description
UPLO	[in]	character*1	= 'U': Upper triangle of A is stored; = 'L': Lower triangle of A is stored.
N	[in]	integer	It is the dimension of A.
A	[in/out]	real*8	A is the N × N input matrix and output of A and array TAU will be used to find reflectors.
LDA	[in]	integer	LDA is the leading dimension of array A which is N.
D	[out]	real*8	D is an array of dimension N which contains diagonal elements after tridiagonalization.
E	[out]	real*8	E is an array of dimension N−1 which contains identical off-diagonal elements after tridiagonalization.
TAU	[out]	real*8	TAU is an array of dimension N−1 and it stores scalars which will be helpful in finding reflectors.
LWORK	[in]	integer	LWORK should be greater than or equal to N.

Argument	In/Out	Type	Description
WORK	[ws/out]	real*8	WORK is an array of dimension equal to LWORK. On exit, if INFO = 0, WORK(0) returns the optimal LWORK.
INFO	[out]	integer	This parameter is for crosschecking the code. If INFO = 0, successful exit.

1.14.10   DORGTR

Syntax

```
call DORGTR( UPLO, N, A, LDA, TAU, WORK, LWORK, INFO )
```

Purpose

DORGTR generates a real orthogonal matrix which transforms the given matrix into a tridiagonal form. DSYTRD and DORGTR has been collectively used for tridiagonalization. The output of DSYTRD will be fed into DORGTR which will finally result in the orthogonal matrix.

Arguments

Argument	In/Out	Type	Description
UPLO	[in]	character*1	= 'U': Upper triangle of A contains elementary reflectors from DSYTRD; = 'L': Lower triangle of A contains elementary reflectors from DSYTRD.
N	[in]	integer	It is the dimension of A
A	[in/out]	real*8	Feed the output of DSYTRD as A which is the N × N input matrix and on exit it will be an orthogonal matrix O.
LDA	[in]	integer	LDA is the leading dimension of array A which is N.
TAU	[in]	real*8	TAU is an array of dimension N−1 which is returned by DSYTRD.
LWORK	[in]	integer	LWORK should be greater than or equal to N.
WORK	[ws/out]	real*8	WORK is an array of dimension equal to LWORK. On exit, if INFO = 0, WORK(0) returns the optimal LWORK.
INFO	[out]	integer	This parameter is for crosschecking the code. If INFO = 0, successful exit.

1.14.11   ZHETRD

Syntax

```
call DSYTRD( UPLO, N, A, LDA, D, E, TAU, WORK, LWORK, INFO )
```

Purpose

ZHETRD reduces the complex Hermitian matrix A into a real symmetric tridiagonal matrix T by an unitary transformation.

$$U \, A \, U^{\dagger} = T \tag{1.3}$$

Arguments

Argument	In/Out	Type	Description
UPLO	[in]	character*1	= 'U': Upper triangle of A is stored; = 'L': Lower triangle of A is stored.
N	[in]	integer	It is the dimension of A.
A	[in/out]	complex*16	A is the N × N input matrix and output of A and array TAU will be used to find reflectors.
LDA	[in]	integer	LDA is the leading dimension of array A which is N.
D	[out]	real*8	D is an array of dimension N which contains the diagonal elements after tridiagonalization.
E	[out]	real*8	E is an array of dimension N−1 which contains identical off-diagonal elements after tridiagonalization.
TAU	[out]	complex*16	TAU is an array of dimension N−1 and it stores scalar which will be helpful in finding reflectors.
LWORK	[in]	integer	LWORK should be greater than or equal to N.
WORK	[ws/out]	complex*16	WORK is an array of dimension equal to LWORK. On exit, if INFO = 0, WORK(0) returns the optimal LWORK.
INFO	[out]	integer	This parameter is for crosschecking the code. If INFO = 0, successful exit.

1.14.12   ZUNGTR

Syntax

```
call ZUNGTR( UPLO, N, A, LDA, TAU, WORK, LWORK, INFO )
```

Purpose

ZUNGTR generates a complex unitary matrix which transforms the given matrix into tridiagonal form. ZHETRD and ZUNGTR has been collectively used for tridiagonalization. The output of ZHETRD will be fed into ZUNGTR which will finally result in the unitary matrix.

Arguments

Argument	In/Out	Type	Description
UPLO	[in]	character*1	= 'U': Upper triangle of A contains elementary reflectors from ZHETRD; = 'L': Lower triangle of A contains elementary reflectors from ZHETRD.
N	[in]	integer	It is the dimension of A.
A	[in/out]	complex*16	Feed the output of ZHETRD as A which is the N × N input matrix and on exit, it will be the unitary matrix U.
LDA	[in]	integer	LDA is the leading dimension of array A which is N.
TAU	[in]	complex*16	TAU is an array of dimension N−1 which is returned by ZHETRD.
LWORK	[in]	integer	LWORK should be greater than or equal to N.
WORK	[ws/out]	complex*16	WORK is an array of dimension equal to LWORK. On exit, if INFO = 0, WORK(0) returns the optimal LWORK.
INFO	[out]	integer	This parameter is for crosschecking the code. If INFO = 0, successful exit.

1.14.13 DGETRF

Syntax

```
call DGETRF( M, N, A, LDA, IPIV, INFO )
```

Purpose

DGETRF computes the LU factorization of the $M \times N$ real matrix A using partial pivoting with row interchanges.

Arguments

Argument	In/Out	Type	Description
M	[in]	integer	It is the number of rows in A.
N	[in]	integer	It is the number of columns in A.
A	[in/out]	real*8	A is the M × N input matrix and on exit the factors L and U are stored.
LDA	[in]	integer	LDA is the leading dimension of array A which is M.
IPIV	[out]	integer	IPIV is an array of dimension min(M,N), the pivot indices; for $1 \le i \le$ min(M,N), row i of the matrix was interchanged with row IPIV(i).
INFO	[out]	integer	This parameter is for crosschecking the code. If INFO = 0, successful exit.

1.14.14   DGETRI

Syntax

```
call DGETRI( N, A, LDA, IPIV, WORK, LWORK, INFO )
```

Purpose

DGETRI computes the inverse of a non-singular real matrix using LU factorization as computed by DGETRF.

Arguments

Argument	In/Out	Type	Description
N	[in]	integer	It is the dimension of A.
A	[in/out]	real*8	Feed the output of DGETRF as A which is the M × N input matrix and on exit, it will store the inverse of A provided, INFO = 0.
LDA	[in]	integer	LDA is the leading dimension of array A which is N.
IPIV	[in]	integer	IPIV is an array of dimension N, the pivot indices from DGETRF; for $1 \le i \le$ N, row i of the matrix was interchanged with row IPIV(i).
LWORK	[in]	integer	LWORK should be greater than or equal to N.

Argument	In/Out	Type	Description
WORK	[ws/out]	real*8	WORK is an array of dimension equal to LWORK. On exit, if INFO = 0, WORK(0) returns the optimal LWORK.
INFO	[out]	integer	This parameter is for crosschecking the code. If INFO = 0, successful exit.

1.14.15   ZGETRF

Syntax

```
call ZGETRF( M, N, A, LDA, IPIV, INFO )
```

Purpose

ZGETRF computes the LU factorization of the $M \times N$ general complex matrix A using partial pivoting with row interchanges.

Arguments

Argument	In/Out	Type	Description
M	[in]	integer	It is the number of rows in A.
N	[in]	integer	It is the number of columns in A.
A	[in/out]	complex*16	A is the M × N input matrix and on exit the factors L and U are stored.
LDA	[in]	integer	LDA is the leading dimension of array A which is M.
IPIV	[out]	integer	IPIV is an array of dimension min(M,N), the pivot indices; for $1 \le i \le$ min(M,N), row i of the matrix was interchanged with row IPIV(i).
INFO	[out]	integer	This parameter is for crosschecking the code. If INFO = 0, successful exit.

1.14.16   ZGETRI

Syntax

```
call ZGETRI( N, A, LDA, IPIV, WORK, LWORK, INFO )
```

Purpose

ZGETRI computes the inverse of a complex matrix using LU factorization as computed by ZGETRF.

Arguments

Argument	In/Out	Type	Description
N	[in]	integer	It is the dimension of A.
A	[in/out]	complex*16	Feed the output of ZGETRF as A which is the $N \times N$ input matrix and on exit, it will store the inverse of A provided, INFO = 0.
LDA	[in]	integer	LDA is the leading dimension of array A which is N.
IPIV	[in]	integer	IPIV is an array of dimension N, the pivot indices from ZGETRF; for $1 \leq i \leq N$, row i of the matrix was interchanged with row IPIV(i).
LWORK	[in]	integer	LWORK should be greater than or equal to N.
WORK	[ws/out]	complex*16	WORK is an array of dimension equal to LWORK. On exit, if INFO = 0, WORK(0) returns the optimal LWORK.
INFO	[out]	integer	This parameter is for crosschecking the code. If INFO = 0, successful exit.

1.14.17   DGESVD

Syntax

```
SUBROUTINE DGESVD( JOBU, JOBVT, M, N, A, LDA, S, U, LDU, VT, LDVT, WORK,
     LWORK, INFO )
```

Purpose

DGESVD computes the singular value decomposition (SVD) of a real $M \times N$ matrix A

$$A = U_{M \times M} S_{M \times N} V_{N \times N}^{T} \tag{1.4}$$

Arguments

Argument	In/Out	Type	Description
JOBU	[in]	CHARACTER*1	Specifies options for computing all or part of the matrix U: = 'A': all M columns of U are returned in the array U; = 'S': the first min(M,N) columns of U (the left singular vectors) are returned in the array U;

Argument	In/Out	Type	Description
			= 'O': the first min(M,N) columns of U (the left singular vectors) are overwritten on the array A; = 'N': no columns of U (no left singular vectors) are computed.
JOBVT	[in]	CHARACTER*1	Specifies options for computing all or part of the matrix V^T = 'A': all N rows of V^T are returned in the array VT; = 'S': the first min(M,N) rows of V^T (the right singular vectors) are returned in the array VT; = 'O': the first min(M,N) rows of V^T (the right singular vectors) are overwritten on the array A; = 'N': no rows of V^T (no right singular vectors) are computed.
M	[in]	integer	It is the number of rows in A.
N	[in]	integer	It is the number of columns in A.
A	[in/out]	real*8	A is M × N input matrix and on exit A will be over written.
LDA	[in]	integer	LDA is the leading dimension of array A which is M.
S	[out]	real*8	S is min(M,N) dimensional array which contains the singular values in ascending order.
U	[out]	real*8	U is an array of dimension LDU × M. If JOBU = 'A', U contains the M × M orthogonal matrix U; if JOBU = 'S', U contains the first min(M,N) columns of U; if JOBU = 'N' or 'O', U is not referenced.
LDU	[in]	integer	LDU is the leading dimension of array U. LDU ≥ 1; if JOBU = 'S' or 'A', LDU ≥ M.
VT	[out]	real*8	VT is an array of dimension LDVT × N. If JOBVT = 'A', U contains the N × N orthogonal matrix V^T; if JOBVT = 'S', VT contains the first min(M,N) columns of V^T; if JOBVT = 'N' or 'O', U is not referenced.
LDVT	[in]	integer	LDVT is a leading dimension of array VT. LDU ≥ 1; if JOBVT = 'S' or 'A', LDVT ≥ N.

Argument	In/Out	Type	Description
LWORK	[in]	integer	LWORK should be greater than or equal to max(1, 3*min(M,N) + max(M,N), 5*min(M,N)).
WORK	[ws/out]	real*8	WORK is an array of dimension equal to LWORK. On exit, if INFO = 0, WORK(0) returns the optimal LWORK.
INFO	[out]	integer	This parameter is for crosschecking the code. If INFO = 0, successful exit.

1.14.18   ZGESVD

Syntax

```
SUBROUTINE ZGESVD( JOBU, JOBVT, M, N, A, LDA, S, U, LDU, VT, LDVT, WORK,
    LWORK, RWORK, INFO )
```

Purpose

ZGESVD computes the singular value decomposition (SVD) of a complex $M \times N$ matrix A

$$A = U_{M \times M} S_{M \times N} V^{\dagger}_{N \times N} \tag{1.5}$$

Arguments

Argument	In/Out	Type	Description
JOBU	[in]	CHARACTER*1	Specifies options for computing all or part of the matrix U: = 'A': all M columns of U are returned in array U; = 'S': the first min(M,N) columns of U (the left singular vectors) are returned in the array U; = 'O': the first min(M,N) columns of U (the left singular vectors) are overwritten on the array A; = 'N': no columns of U (no left singular vectors) are computed.
JOBVT	[in]	CHARACTER*1	Specifies options for computing all or part of the matrix $V^{\dagger}$ = 'A': all N rows of $V^{\dagger}$ are returned in the array VT; = 'S': the first min(M,N) rows of $V^{\dagger}$ (the right singular vectors) are returned in the array VT;

Argument	In/Out	Type	Description
			= 'O': the first min(M,N) rows of $V^\dagger$ (the right singular vectors) are overwritten on the array A; = 'N': no rows of $V^\dagger$ (no right singular vectors) are computed.
M	[in]	integer	It is the number of rows in A.
N	[in]	integer	It is the number of columns in A.
A	[in/out]	complex*16	A is the M $\times$ N input matrix and on exit A will be over written.
LDA	[in]	integer	LDA is the leading dimension of array A which is M.
S	[out]	real*8	S is min(M,N) dimensional array which contains the singular values in ascending order.
U	[out]	complex*16	U is the array of dimension LDU $\times$ M. If JOBU = 'A', U contains the M $\times$ M unitary matrix U; if JOBU = 'S', U contains the first min(M,N) columns of U; if JOBU = 'N' or 'O', U is not referenced.
LDU	[in]	integer	LDU is a leading dimension of array U. LDU $>=$ 1; if JOBU = 'S' or 'A', LDU $\geq$ M.
VT	[out]	complex*16	VT is the array of dimension LDVT $\times$ N. If JOBVT = 'A', U contains the N $\times$ N unitary matrix V^T; if JOBVT = 'S', VT contains the first min(M,N) columns of V^T; if JOBVT = 'N' or 'O', U is not referenced.
LDVT	[in]	integer	LDVT is a leading dimension of array VT. LDU $\geq$ 1; if JOBVT = 'S' or 'A', LDVT $\geq$ N.
LWORK	[in]	integer	LWORK should be greater than or equal to max(1, 2*min(M,N)+max(M,N)).
WORK	[ws/out]	complex*16	WORK is an array of the dimension equal to LWORK. On exit, if INFO = 0, WORK(0) returns the optimal LWORK.
RWORK	[ws]	real*8	RWORK is an array of dimension 5*min(M,N).
INFO	[out]	integer	This parameter is for crosschecking the code. If INFO = 0, successful exit.

1.15 Compilation and execution

Once the program is written, we have to compile the program and execute it, for that, FORTRAN has a specific set of procedures which we explain here. We use two compilers here as we have already seen - one is gfortran (open source) and other, the ifort (provided by INTEL Math Kernel Libraries). The compilation is done via the terminal in the linux operating system.

1. To compile a program which does not use any external libraries,

   - **gfortran myprogram.f90**
   - **ifort myprogram.f90**

2. To compile a program which uses external libraries like LAPACK and BLAS,

   - **gfortran myprogram.f90 -llapack -lblas**
   - **ifort myprogram.f90 -mkl**

3. For linking more than one programs (for e.g. myprogram1.f90 and myprogram2.f90) together, we adopt,

   - **gfortran myprogram1.f90 myprogram2.f90 -llapack -lblas**
   - **ifort myprogram1.f90 myprogram2.f90 -mkl**

4. After the compilation of the program as above, the compiler will generate an executable file named 'a.out' in the same directory where the program is present. We will get the output from the program only if we execute this aforesaid file. For doing this, we have to invoke the command as below.
   ./a.out

The results of this execution will be available to the user either as a text file or as an on screen data depending on the user's choice in the way in which the user has written file.

1.16 Computation time

In many situations, the user needs to find the computation time for some block of the code or maybe the whole code, there is an intrinsic function available in FORTRAN for that too, which is known as **call system_clock()**. An example is given below.

```
program example
integer::t1,t2,clock_max,factn
real*8::clock_rate ,time
call system_clock ( t1, clock_rate, clock_max )
.............................................
......program block under inspection.....
.............................................
call system_clock ( t2, clock_rate, clock_max )
time= real( t2 - t1 )/clock_rate
end program
```

The system clock function as described by the above code can be used to find the computational time of the entire code or a part of the code. The variable 'time' gives the exact time of computation. Note that all the times are calculated in seconds.

2

Nuts and Bolts of Quantum Mathematics

In the last chapter, we have gained an understanding of the FORTRAN 90 language and its features. In this chapter we will proceed with the development of numerical subroutines for some basic operations of quantum mechanics which will be helpful in developing other complicated subroutines. We have assumed that the reader will have a good grasp of basic quantum mechanics [6–10] at the level of post graduation in order to understand the terminologies and the subroutines in this chapter as well as the entire book without much difficulty. The level of difficulty of the subroutines will increase as we travel through the chapters. Here we have used the standard quantum mechanics convention for all the operations; for example, any pure quantum state is described by a ket vector $| \ \rangle$ and its corresponding dual by the bra vector which is $\langle \ |$. In this chapter we have any vector $|v\rangle$ as a $n \times 1$ column matrix, also called "ket v" which can be written explicitly as

$$|v\rangle = \begin{pmatrix} v_1 \\ v_2 \\ \vdots \\ v_i \\ \vdots \\ v_n \end{pmatrix}. \tag{2.1}$$

Also, note that the Hermitian conjugate of $|v\rangle$ is a $1 \times n$ row matrix called "bra v" written as $\langle v|$ which has an explicit form given by

$$\begin{pmatrix} v_1^* & v_2^* & \cdots & v_i^* & \cdots & v_n^* \end{pmatrix}. \tag{2.2}$$

In Eqs. [2.1] and [2.2], the entries of the matrix namely the v_i's can either be real or complex corresponding to a real vector and a complex vector, respectively. Any one qubit state $|\psi\rangle$ is written as a linear combination of two bits namely $|0\rangle$ and $|1\rangle$ (eigenstates of the Pauli σ_z matrix) as $|\psi\rangle = \alpha|0\rangle + \beta|1\rangle$ such that $\alpha, \beta \in \mathbb{C}$ and $|\alpha|^2 + |\beta|^2 = 1$. The corresponding dual of $|\psi\rangle$ which is the state $\langle \psi|$ is given by $\alpha^*\langle 0| + \beta^*\langle 1|$. It is well known that, one qubit states are very fundamental in quantum information [11] as multiqubit states can be built using them. Any physically observable or measurable quantity in quantum mechanics is represented as a Hermitian operator and thereby it has a matrix representation in a given basis. Keeping these in mind, we have developed recipes which involve vector-vector and matrix-matrix operations [12, 13]. Not only that, since these multiqubit quantum states can be represented as binary strings, much simplifications can be done in programming by a clever interconversion of binary and decimal entities. To elucidate more, any operation involving a product of $2^N \times 2^N$ matrix with a $2^N \times 1$ column vector, just boils down to an operation of the given operator on the bits rather than the whole 2^N qubit state. The power of such a simplification can be realized when we construct Hamiltonian matrix.

2.1   Inner product between two vectors

Let there be two n-dimensional vectors $|v\rangle$ and $|w\rangle$ in a Hilbert space $\mathcal{H}$ of dimension n defining a physical system; here $|v\rangle$ and $|w\rangle$ are represented by $n \times 1$ column matrices. The inner product between these two vectors can be given in terms of matrix elements as

$$c = \langle v|w\rangle = \sum_{i=1}^{n} v_i^* w_i, \tag{2.3}$$

where v_i and w_i are the ith entries or matrix elements of the vectors $|v\rangle$ and $|w\rangle$, respectively. The inner product is a very important operation in the space $\mathcal{H}$ as it is extensively used to understand properties such as orthogonality and normalization of vectors as these are ubiquitous in quantum mechanics.

2.1.1   For real vectors

```
subroutine IPVR(v1,v2,n,c)
```

Parameters

In/Out	Argument	Description
[in]	n	n (integer) is the dimension of the vectors between which we are calculating the inner product
[in]	v1	v1 is a real*8 array of dimension (0:n−1)
[in]	v2	v2 is a real*8 array of dimension (0:n−1)
[out]	c	c (real*8) is the inner product between vectors v1 and v2

Implementation

```
subroutine IPVR(v1,v2,n,c)
implicit none
integer::n,i
real*8::v1(0:n-1),v2(0:n-1),c
c=0.0d0
do i=0,n-1
    c=c+v1(i)*v2(i)
end do
end subroutine
```

Example

In this example we are finding the inner product between the vectors $v_1 = (1, 2, 3, 4)^T$ and $v_2 = (2, 3, 1, -1)^T$.

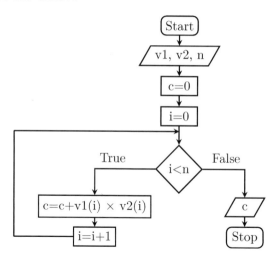

FIGURE 2.1: Flowchart of IPVR

```fortran
program example
implicit none
real*8::v1(0:3),v2(0:3),c
v1=(/1.0,2.0,3.0,4.0/)
v2=(/2.0,3.0,1.0,-1.0/)
call IPVR(v1,v2,4,c)
print*,c
end program
```

Prints, to standard output which is the inner product between two vectors.

```
7.0000000000000000
```

2.1.2 For complex vectors

```fortran
subroutine IPVC(v1,v2,n,c)
```

Parameters

In/Out	Argument	Description
[in]	n	n (integer) is the dimension of the vectors between which we are calculating the inner product
[in]	v1	v1 is a complex*16 array of dimension $(0:n-1)$
[in]	v2	v2 is a complex*16 array of dimension $(0:n-1)$
[out]	c	c (complex*16) is the inner product between vectors v1 and v2

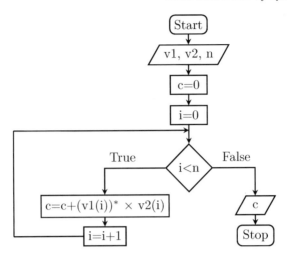

FIGURE 2.2: Flowchart of IPVC

Implementation

```
subroutine IPVC(v1,v2,n,c)
implicit none
integer::n,i
complex*16::v1(0:n-1),v2(0:n-1),c
c=0.0d0
do i=0,n-1
    c=c+dconjg(v1(i))*v2(i)
end do
end subroutine
```

Example

In this example we are finding the inner product between the vectors $v_1 = (1, 1 - i, 2 + 3i)^T$ and $v_2 = (2, 3 - i, 2 - 3i)^T$.

```
program example
implicit none
complex*16::v1(0:2),v2(0:2),c
v1(0)=dcmplx(1,0)
v1(1)=dcmplx(1,-1)
v1(2)=dcmplx(2,3)
v2(0)=dcmplx(2,0)
v2(1)=dcmplx(3,-1)
v2(2)=dcmplx(2,-3)
call IPVC(v1,v2,3,c)
print*,c
end program
```

Prints, to standard output which is the inner product between the two vectors.

```
(1.0000000000000000,-10.000000000000000)
```

2.2 Norm of a vector

For a given n-dimensional vector $|v\rangle$ in the Hilbert space, using the inner product of the vectors as given by Eq. [2.3], the Euclidean norm of the vector $|v\rangle$ can be defined as

$$||v|| = \sqrt{\langle v|v\rangle} = \sqrt{\sum_{i=1}^{n} v_i^* v_i} = \sqrt{\sum_{i=1}^{n} |v_i|^2}, \tag{2.4}$$

the norm is a very important number in the Hilbert space, it is used as a measure to define the length of the abstract vector $|v\rangle$.

2.2.1 For a real vector

```
subroutine NORMVR(v,n,norm)
```

Parameters

In/Out	Argument	Description
[in]	n	n (integer) is the dimension of the vector v
[in]	v	v is a real*8 array of dimension (0:n−1)
[out]	norm	norm (real*8) is the norm of the vector v

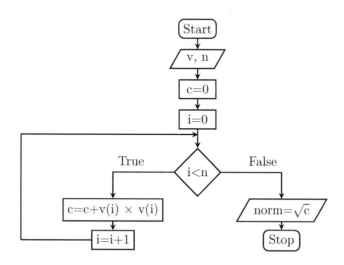

FIGURE 2.3: Flowchart of NORMVR

Implementation

```
subroutine NORMVR(v,n,norm)
implicit none
integer::n,i
real*8::v(0:n-1),c,norm
c=0.0d0
do i=0,n-1
    c=c+v(i)*v(i)
end do
norm=dsqrt(c)
end subroutine
```

Example

In this example we are finding the norm of the vector $v = (1, 2, 3, 4)^T$.

```
program example
implicit none
real*8::v(0:3),norm
v=(/1.0,2.0,3.0,4.0/)
call NORMVR(v,4,norm)
print*,norm
end program
```

Prints, to standard output which is the norm of the vector.

```
5.4772255750516612
```

2.2.2   For a complex vector

```
subroutine NORMVC(v,n,norm)
```

Parameters

In/Out	Argument	Description
[in]	n	n (integer) is the dimension of the vector v
[in]	v	v is a complex*16 array of dimension $(0:n-1)$
[out]	norm	norm (real*8) is the norm of the vector v

Implementation

```
subroutine NORMVC(v,n,norm)
implicit none
integer::n,i
complex*16::v(0:n-1),c
```

```
real*8::norm
c=0.0d0
do i=0,n-1
    c=c+conjg(v(i))*v(i)
end do
norm=dsqrt(abs(c))
end subroutine
```

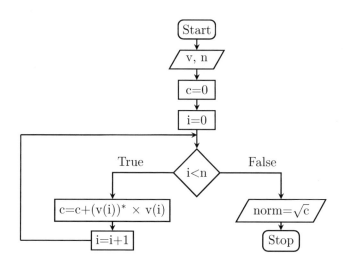

FIGURE 2.4: Flowchart of NORMVC

Example

In this example we are finding the norm of the vector $v = (1, 1 - i, 2 + 3i)^T$.

```
program example
implicit none
complex*16::v(0:2)
real*8::norm
v(0)=dcmplx(1,0)
v(1)=dcmplx(1,-1)
v(2)=dcmplx(2,3)
call NORMVC(v,3,norm)
print*,norm
end program
```

Prints, to standard output which is the norm of the vector.

4.0000000000000000

2.3   Normalization of a vector

Once the norm is found using Eq. [2.4], we can proceed to normalize the n-dimensional vector $|v\rangle$ to obtain the normalized vector $|v'\rangle$ as

$$|v'\rangle = \frac{|v\rangle}{||v||},\qquad(2.5)$$

normalization is an important operation which is done on quantum states so as to make the length of the vector representing the state to be unity, that is to say, even though $\langle v|v\rangle$ may not be unity in general, it is guaranteed that $\langle v'|v'\rangle$ is 1.

2.3.1   For a real vector

subroutine NORMALIZEVR(v,n)

Parameters

In/Out	Argument	Description
[in]	n	n (integer) is the dimension of the vector v
[in/out]	v	v is a real*8 array of dimension (0:n−1), on exit, it will be overwritten by the normalized vector

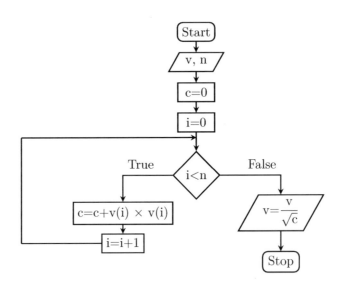

FIGURE 2.5: Flowchart of NORMALIZEVR

Implementation

```fortran
subroutine NORMALIZEVR(v,n)
implicit none
integer::n,i
real*8::v(0:n-1),c
c=0.0d0
do i=0,n-1
    c=c+v(i)*v(i)
end do
v=v/dsqrt(c)
end subroutine
```

Example

In this example we are normalizing the vector $v = (1, 2, 3, 4)^T$.

```fortran
program example
real*8::v(0:3)
integer::i
v=(/1.0,2.0,3.0,4.0/)
call NORMALIZEVR(v,4)
do i=0,3,1
    print*,v(i)
end do
end program
```

Prints, to standard output which is the normalized vector.

```
0.18257418583505536
0.36514837167011072
0.54772255750516607
0.73029674334022143
```

2.3.2   For a complex vector

```fortran
subroutine NORMALIZEVC(v,n)
```

Parameters

In/Out	Argument	Description
[in]	n	n (integer) is the dimension of the vector v
[in]	v	v is a complex*16 array of dimension (0:n−1), on exit, it will be overwritten by the normalized vector

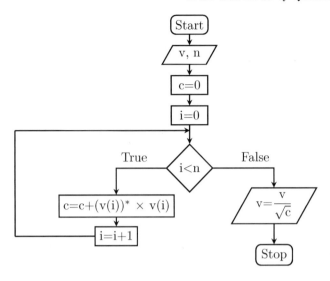

FIGURE 2.6: Flowchart of NORMALIZEVC

Implementation

```
subroutine NORMALIZEVC(v,n)
implicit none
integer::n,i
complex*16::v(0:n-1),c
c=0.0d0
do i=0,n-1
     c=c+conjg(v(i))*v(i)
end do
v=v/dsqrt(abs(c))
end subroutine
```

Example

In this example we are normalizing the vector $v = (1, 1-i, 2+3i)^T$.

```
program example
implicit none
complex*16::v(0:2)
integer::i
v(0)=dcmplx(1,0)
v(1)=dcmplx(1,-1)
v(2)=dcmplx(2,3)
call NORMALIZEVC(v,3)
do i=0,2,1
    print*,v(i)
end do
end program
```

Prints, to standard output which is the normalized vector.

```
(0.25000000000000000,0.0000000000000000)
(0.25000000000000000,-0.25000000000000000)
(0.50000000000000000,0.75000000000000000)
```

2.4   Outer product of a vector

If we have two vectors $|v\rangle$ and $|w\rangle$ of dimensions n_1 and n_2, respectively, the outer product between them is an operator which has the matrix representation A having the dimension $n_1 \times n_2$ given by,

$$A = |v\rangle\langle w|, \qquad (2.6)$$

and the ijth matrix element of A is given by,

$$A_{ij} = \langle i|A|j\rangle = v_i w_j^*. \qquad (2.7)$$

The outer product is an important entity in quantum mechanics which has many uses, especially used in the representation of projection operators, density matrices, and so on. When $|v\rangle = |w\rangle$, it represents the projection operator.

2.4.1   For real vectors

```
subroutine OPVR(v,w,n1,n2,A)
```

Parameters

In/Out	Argument	Description
[in]	n1	n1 (integer) is the dimension of the vector v
[in]	n2	n2 (integer) is the dimension of the vector w
[in]	v	v is a real*8 array of dimension (0:n1−1)
[in]	w	w is a real*8 array of dimension (0:n2−1)
[out]	A	A is a real*8 array of dimension (0:n1−1,0:n2−1), which is the outer product

Implementation

```
subroutine OPVR(v,w,n1,n2,A)
implicit none
integer::n1,n2,i,j
real*8::v(0:n1-1),w(0:n2-1),A(0:n1-1,0:n2-1)
do i=0,n1-1
   do j=0,n2-1
      A(i,j)=v(i)*w(j)
```

```
      end do
end do
end subroutine
```

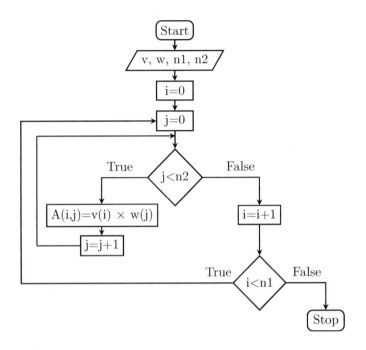

FIGURE 2.7: Flowchart of OPVR

Example

In this example we are finding the outer product between vectors $v_1 = (1,2)^T$ and $v_2 = (2,3,1)^T$.

```
program example
implicit none
real*8::v1(0:1),v2(0:2),A(0:1,0:2)
integer::i
v1=(/1.0,2.0/)
v2=(/2.0,3.0,1.0/)
call OPVR(v1,v2,2,3,A)
do i=0,1,1
write(111,*)A(i,:)
end do
end program
```

The file "fort.111" will contain the required outer product of vectors $v1$ and $v2$.

```
2.0000000000000000  3.0000000000000000  1.0000000000000000
4.0000000000000000  6.0000000000000000  2.0000000000000000
```

2.4.2 For complex vectors

```
subroutine OPVC(v,w,n1,n2,A)
```

Parameters

In/Out	Argument	Description
[in]	n1	n1 (integer) is the dimension of the vector v
[in]	n2	n2 (integer) is the dimension of the vector w
[in]	v	v is a complex*16 array of dimension (0:n1−1)
[in]	w	w is a complex*16 array of dimension (0:n2−1)
[out]	A	A is a complex*16 array of dimension (0:n1−1,0:n2−1), which is the outer product

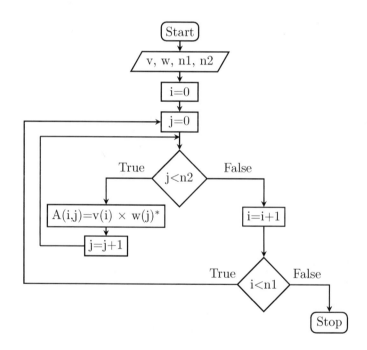

FIGURE 2.8: Flowchart of OPVC

Implementation

```
subroutine OPVC(v,w,n1,n2,A)
implicit none
integer::n1,n2,i,j
complex*16::v(0:n1-1),w(0:n2-1),A(0:n1-1,0:n2-1)
do i=0,n1-1
```

```
        do j=0,n2-1
            A(i,j)=v(i)*conjg(w(j))
        end do
end do
end subroutine
```

Example

In this example we are finding the outer product between vectors $v_1 = (1, 1 - i)^T$ and $v_2 = (2, 3 - i)^T$.

```
program example
implicit none
complex*16::v1(0:1),v2(0:1),A(0:1,0:1)
integer::i,j
v1(0)=dcmplx(1,0)
v1(1)=dcmplx(1,-1)
v2(0)=dcmplx(2,0)
v2(1)=dcmplx(3,-1)
call OPVC(v1,v2,2,2,A)
do i=0,1
    do j=0,1
        write(111,*) A(i,j)
    end do
end do
end program
```

The file "fort.111" will contain the required outer product of vectors $v1$ and $v2$.

```
(2.0000000000000000,0.0000000000000000)
(3.0000000000000000,1.0000000000000000)
(2.0000000000000000,-2.0000000000000000)
(4.0000000000000000,-2.0000000000000000)
```

2.5   Matrix multiplication

Let there are two matrices A and B of dimensions $m \times n$ and $n \times p$, respectively. The matrix multiplication between A and B resulting in a matrix C of dimension $m \times p$ can be given by

$$C = AB. \tag{2.8}$$

However, writing it explicitly in terms of the matrix elements for better clarity, we have

$$C_{ij} = \sum_{k=1}^{n} A_{ik} B_{kj}, \tag{2.9}$$

with C_{ij} being any matrix element of the product AB. However, in FORTRAN 90 and above, we have an intrinsic function called MATMUL which can perform matrix multiplications with ease. The above matrix multiplication can be simply written as $C = \text{MATMUL}(A, B)$.

Here we have not used MATMUL function as it eclipses certain internal details on matrix multiplication looping structures which a beginner should know. As we know, matrix multiplications occur in any field, be it the sciences, humanities or economics. The flowcharts for real and complex matrices are the same.

2.5.1 For real matrices

---
subroutine MMULMR(A,B,m,n,p,C)

Parameters

In/Out	Argument	Description
[in]	m	m (integer) is the number of rows in the matrix A
[in]	n	n (integer) is the number of columns in the matrix A or the number of rows in the matrix B
[in]	p	p (integer) is the number of columns in the matrix B
[in]	A	A is a real*8 array of dimension (0:m−1,0:n−1)
[in]	B	B is a real*8 array of dimension (0:n−1,0:p−1)
[out]	C	C is a real*8 array of dimension (0:m−1,0:p−1) which is the product of matrices A and B

Implementation

```
subroutine MMULMR(A,B,m,n,p,C)
implicit none
integer::n,m,l,i,j,k,p
real*8::A(0:m-1,0:n-1),B(0:n-1,0:p-1),C(0:m-1,0:p-1),temp
do i=0,m-1,1
    do j=0,p-1,1
        temp = 0.0d0
        do k=0,n-1,1
            temp=temp+A(i,k)*B(k,j)
        end do
        C(i,j) = temp
    end do
end do
end subroutine
```

Example

In this example we are multiplying two matrices as given below:

$$A = \begin{pmatrix} 1 & 2 \\ 2 & -1 \end{pmatrix}, \quad B = \begin{pmatrix} 1 & 2 \\ 3 & 1 \end{pmatrix}. \tag{2.10}$$

```
program example
implicit none
real*8::A(0:1,0:1),B(0:1,0:1),C(0:1,0:1)
integer::i
A(0,:)=(/1.0,2.0/)
A(1,:)=(/2.0,-1.0/)
B(0,:)=(/1.0,2.0/)
B(1,:)=(/3.0,1.0/)
call MMULMR(A,B,2,2,2,C)
do i=0,1,1
write(111,*) C(i,:)
end do
end program
```

The file "fort.111" will contain the product of matrix A and B.

| 7.0000000000000000 | 4.0000000000000000 |
| -1.0000000000000000 | 3.0000000000000000 |

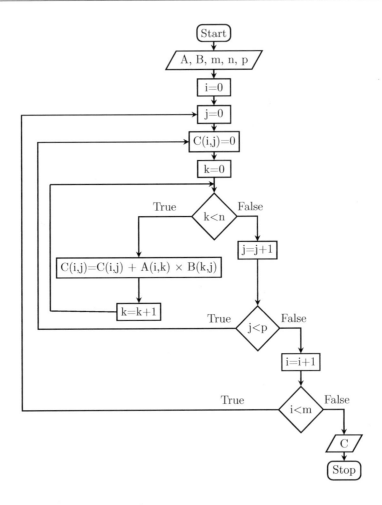

FIGURE 2.9: Flowchart of MMULMR and MMULMC

2.5.2 For complex matrices

```
subroutine MMULMC(A,B,m,n,p,C)
```

Parameters

In/Out	Argument	Description
[in]	m	m (integer) is the number of rows in the matrix A
[in]	n	n (integer) is the number of columns in the matrix A or the number of rows in the matrix B
[in]	p	p (integer) is the number of columns in the matrix B
[in]	A	A is a complex*16 array of dimension $(0:m-1,0:n-1)$
[in]	B	B is a complex*16 array of dimension $(0:n-1,0:p-1)$
[out]	C	C is a complex*16 array of dimension $(0:m-1,0:p-1)$ which is the product of matrices A and B

Implementation

```
subroutine MMULMC(A,B,m,n,p,C)
implicit none
integer::n,m,l,i,j,k,p
complex*16::A(0:m-1,0:n-1),B(0:n-1,0:p-1),C(0:m-1,0:p-1),temp
do i=0,m-1,1
    do j=0,p-1,1
        temp = 0.0d0
        do k=0,n-1,1
            temp=temp+A(i,k)*B(k,j)
        end do
        C(i,j) = temp
    end do
end do
end subroutine
```

Example

In this example we are multiplying two matrices as given below:

$$A = \begin{pmatrix} 1 & 1-i \\ 2+3i & 4 \end{pmatrix}, \quad B = \begin{pmatrix} 1+i & 1+i \\ 3-3i & 4-i \end{pmatrix}. \tag{2.11}$$

```
program example
implicit none
complex*16::A(0:1,0:1),B(0:1,0:1),C(0:1,0:1)
```

```
integer::i,j
A(0,0)=dcmplx(1,0)
A(0,1)=dcmplx(1,-1)
A(1,0)=dcmplx(2,3)
A(1,1)=dcmplx(4,0)
B(0,0)=dcmplx(1,1)
B(0,1)=dcmplx(1,1)
B(1,0)=dcmplx(3,-3)
B(1,1)=dcmplx(4,-1)
call MMULMC(A,B,2,2,2,C)
do i=0,1
    do j=0,1
        write(111,*) C(i,j)
    end do
end do
end program
```

The file "fort.111" will contain the product of matrix A and B.

```
(1.0000000000000000,-5.0000000000000000)
(4.0000000000000000,-4.0000000000000000)
(11.000000000000000,-7.0000000000000000)
(15.000000000000000,1.0000000000000000)
```

2.6   Tensor product between two matrices

Tensor product is also known as direct product or Kronecker product. Tensor products are not restricted to square matrices only but valid even for rectangular matrices equally. Let there be two matrices A and B of the dimensions $m \times n$ and $p \times q$, respectively. The Kronecker product between these two matrices will be a matrix of dimension $mp \times nq$ can be represented as below:

$$A \otimes B = \begin{pmatrix} A_{11}B & A_{12}B & \cdots & A_{1n}B \\ A_{21}B & A_{22}B & \cdots & A_{2n}B \\ \vdots & \vdots & \ddots & \vdots \\ A_{m1}B & A_{m2}B & \cdots & A_{mn}B \end{pmatrix}_{mp \times nq} . \tag{2.12}$$

Note that, each of the elements of the matrix A is algebraically multiplied by the total matrix B. Kronecker products are very useful when we deal with several identical or non-identical physical systems and to find the quantum states representing those systems. For example, any general one-qubit system can be represented by the state $|\phi\rangle = \alpha|0\rangle + \beta|1\rangle$, now, if we have to find the two-qubit state $|\xi\rangle$ corresponding to two one-qubit states, then we can have

$$|\xi\rangle = (\alpha_1|0\rangle + \beta_1|1\rangle)_1 \otimes (\alpha_1|0\rangle + \beta_1|1\rangle)_2. \tag{2.13}$$

The subscripts in the above refer to systems 1 and 2, respectively. Note that Eq. [2.13] can be extended to N qubit states. Not only that, the concept of tensor product can be extended to operators which represent physical observables in quantum mechanics also. The flowcharts for real and complex matrices are same.

2.6.1 For real matrices

subroutine KPMR(m,n,A,p,q,B,C)

Parameters

In/Out	Argument	Description
[in]	m	m (integer) is the number of rows in the matrix A
[in]	n	n (integer) is the number of columns in the matrix A
[in]	A	A is a real*8 array of dimension (0:m−1,0:n−1)
[in]	p	p (integer) is the number of rows in the matrix B
[in]	q	q (integer) is the number of columns in the matrix B
[in]	B	B is a real*8 array of dimension (0:p−1,0:q−1)
[out]	C	C is a real*8 array of dimension (0:m∗p−1,0:n∗q−1), which is the tensor product of matrices A and B

Implementation

```
subroutine KPMR(m,n,A,p,q,B,C)
implicit none
integer::m,n,p,q,i,j,k,l
real*8,dimension(0:m-1,0:n-1)::A
real*8,dimension(0:p-1,0:q-1)::B
real*8,dimension(0:m*p-1,0:n*q-1)::C
do k=0,m-1,1
    do l=0,p-1,1
        do i=0,n-1,1
            do j=0,q-1,1
                C(p*k+l,q*i+j)=A(k,i)*B(l,j)
            end do
        end do
    end do
end do
end subroutine
```

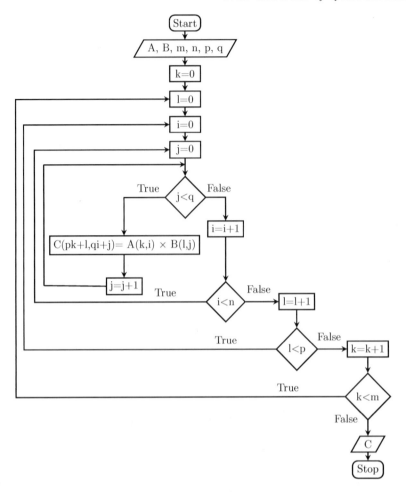

FIGURE 2.10: Flowchart of KPMR and KPMC

Example

In this example we are finding the tensor product between two matrices as given below:

$$A = \begin{pmatrix} 1 \\ 2 \end{pmatrix}, \quad B = \begin{pmatrix} 1 & 2 \\ 3 & 1 \end{pmatrix}. \tag{2.14}$$

```
program example
implicit none
real*8::A(0:1,0:0),B(0:1,0:1),C(0:3,0:1)
integer::i
A(0,0)=1.0
A(1,0)=2.0
B(0,:)=(/1.0,2.0/)
B(1,:)=(/3.0,1.0/)
call KPMR(2,1,A,2,2,B,C)
do i=0,3,1
```

```
write(111,*) C(i,:)
end do
end program
```

The file "fort.111" will contain the tensor product between the matrices A and B.

1.0000000000000000	2.0000000000000000
3.0000000000000000	1.0000000000000000
2.0000000000000000	4.0000000000000000
6.0000000000000000	2.0000000000000000

2.6.2   For complex matrices

```
subroutine KPMC(m,n,A,p,q,B,C)
```

Parameters

In/Out	Argument	Description
[in]	m	m (integer) is the number of rows in the matrix A
[in]	n	n (integer) is the number of columns in the matrix A
[in]	A	A is a complex*16 array of dimension (0:m−1,0:n−1)
[in]	p	p (integer) is the number of rows in the matrix B
[in]	q	q (integer) is the number of columns in the matrix B
[in]	B	B is a complex*16 array of dimension (0:p−1,0:q−1)
[out]	C	C is a complex*16 array of dimension (0:m∗p−1,0:n∗q−1), which is the tensor product of matrices A and B

Implementation

```
subroutine KPMC(m,n,A,p,q,B,C)
implicit none
integer::m,n,p,q,i,j,k,l
complex*16,dimension(0:m-1,0:n-1)::A
complex*16,dimension(0:p-1,0:q-1)::B
complex*16,dimension(0:m*p-1,0:n*q-1)::C
do k=0,m-1,1
    do l=0,p-1,1
        do i=0,n-1,1
            do j=0,q-1,1
                C(p*k+l,q*i+j)=A(k,i)*B(l,j)
            end do
        end do
    end do
end do
```

```
end do
end subroutine
```

Example

In this example we are finding the tensor product between two matrices as given below:

$$A = \begin{pmatrix} 1 \\ 1-i \end{pmatrix}, \quad B = \begin{pmatrix} 1+i \\ 3-3i \end{pmatrix}. \tag{2.15}$$

```
program example
implicit none
complex*16::A(0:1,0:0),B(0:1,0:0),C(0:3,0:0)
integer::i
A(0,0)=dcmplx(1,0)
A(1,0)=dcmplx(1,-1)
B(0,0)=dcmplx(1,1)
B(1,0)=dcmplx(3,-3)
call KPMC(2,1,A,2,1,B,C)
do i=0,3,1
   write(111,*) C(i,:)
end do
end program
```

The file "fort.111" will contain the tensor product between the matrices A and B.

```
(1.0000000000000000,1.0000000000000000)
(3.0000000000000000,-3.0000000000000000)
(2.0000000000000000,0.0000000000000000)
(0.0000000000000000,-6.0000000000000000)
```

2.7   Trace of a matrix

Trace of a matrix is the sum of all its diagonal elements. Let us have an $n \times n$ matrix A. Trace of the matrix A can be defined as

$$tr(A) = \sum_{i=1}^{n} A_{ii}. \tag{2.16}$$

The trace finds its use in quantum mechanics, especially when we deal with density operators ρ which has to satisfy the condition $tr(\rho) = 1$. There are other areas in which trace of a matrix proves to be a useful quantity which we will see as the book evolves. The flowcharts for finding the trace of both real and complex matrices are same.

2.7.1   For a real matrix

```
subroutine TRMR(A,n,trace)
```

Parameters

In/Out	Argument	Description
[in]	n	n (integer) is the dimension of the matrix A
[in]	A	A is a real*8 array of dimension $(0:n-1,0:n-1)$
[out]	trace	trace (real*8) is the trace of the matrix A

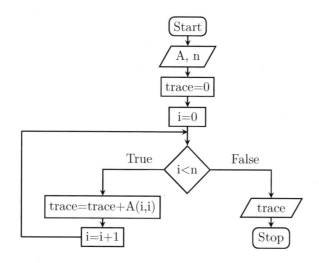

FIGURE 2.11: Flowchart of TRMR and TRMC

Implementation

```
subroutine TRMR(A,n,trace)
integer::i,n
real*8::A(0:n-1,0:n-1),trace
trace=0.0d0
do i=0,n-1
    trace=trace+A(i,i)
end do
end subroutine
```

Example

In this example we are finding the trace of matrix given below:

$$A = \begin{pmatrix} 1 & 2 & 3 \\ 3 & -2 & -5 \\ 7 & -2 & -3 \end{pmatrix}. \tag{2.17}$$

```
program example
implicit none
real*8::A(0:2,0:2),trace
A(0,:)=(/1.0,2.0,3.0/)
A(1,:)=(/3.0,-2.0,-5.0/)
A(2,:)=(/7.0,-2.0,-3.0/)
call TRMR(A,3,trace)
write(111,*) trace
end program
```

The file "fort.111" will contain the trace of matrix A.

```
-4.0000000000000000
```

2.7.2   For a complex matrix

```
subroutine TRMC(A,n,trace)
```

Parameters

In/Out	Argument	Description
[in]	n	n (integer) is the dimension of the matrix A
[in]	A	A is a complex*16 array of dimension $(0{:}n{-}1,0{:}n{-}1)$
[out]	trace	trace (complex*16) is the trace of the matrix A

Implementation

```
subroutine TRMC(A,n,trace)
integer::i,n
complex*16::A(0:n-1,0:n-1),trace
trace=0.0d0
do i=0,n-1
   trace=trace+A(i,i)
end do
end subroutine
```

Example

In this example we are finding trace of matrix given below:

$$A = \begin{pmatrix} 1+i & 1+i \\ 3-3i & 4+i \end{pmatrix}. \tag{2.18}$$

```
program example
implicit none
complex*16::B(0:1,0:1),trace
B(0,0)=dcmplx(1,1)
B(0,1)=dcmplx(1,1)
B(1,0)=dcmplx(3,-3)
B(1,1)=dcmplx(4,1)
call TRMC(B,2,trace)
write(111,*)trace
end program
```

The file "fort.111" will contain the trace of matrix B.

```
(5.0000000000000000,2.0000000000000000)
```

2.8   Commutator between two matrices

In general, it is known that matrices do not commute under multiplication, so it is useful to define the quantity called commutator represented by $[A, B]$ between two matrices and is given by,

$$C = [A, B] = AB - BA, \tag{2.19}$$

if $[A, B] = 0 \Rightarrow AB = BA$, then matrices A and B are said to commute with each other. Commutator arises as a basic principle in quantum mechanics where we start with the canonical commutation relation [14] between the ith and jth component of position and momentum, respectively, given by $[x_i, p_j] = i\hbar\delta_{ij}$ from which the entire edifice of quantum mechanics is built. It is also worth mentioning here that, if two physical observables do not commute, then they cannot be diagonalized by a common set of eigenvectors; this has deep implications for the Heisenberg uncertainty principle, etc. The flowcharts of finding commutator of real and complex matrices are the same.

2.8.1   For real matrices

```
subroutine COMMR(n,A,B,C)
```

Parameters

In/Out	Argument	Description
[in]	n	n (integer) is the dimension of the matrices A and B
[in]	A	A is a real*8 array of dimension (0:n−1,0:n−1)
[in]	B	B is a real*8 array of dimension (0:n−1,0:n−1)
[out]	C	C is a real*8 array of dimension (0:n−1,0:n−1) which is the commutator between matrices A and B

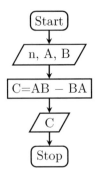

FIGURE 2.12: Flowchart of COMMR and COMMC

Implementation

```
subroutine COMMR(n,A,B,C)
implicit none
integer::n
real*8::A(0:n-1,0:n-1),B(0:n-1,0:n-1),C(0:n-1,0:n-1)
C=matmul(A,B)-matmul(B,A)
end subroutine
```

Example

In this example we are finding the commutator between two matrices as given below:

$$A = \begin{pmatrix} 1 & 2 \\ 2 & -1 \end{pmatrix}, \quad B = \begin{pmatrix} 1 & 2 \\ 3 & 1 \end{pmatrix}. \tag{2.20}$$

```
program example
implicit none
real*8::A(0:1,0:1),B(0:1,0:1),C(0:1,0:1)
integer::i
A(0,:)=(/1.0,2.0/)
A(1,:)=(/2.0,-1.0/)
B(0,:)=(/1.0,2.0/)
B(1,:)=(/3.0,1.0/)
call COMMR(2,A,B,C)
do i=0,1,1
write(111,*) C(i,:)
end do
end program
```

The file "fort.111" will contain the commutator between matrices A and B.

2.0000000000000000	4.0000000000000000
-6.0000000000000000	-2.0000000000000000

2.8.2   For complex matrices

```
subroutine COMMC(n,A,B,C)
```

Parameters

In/Out	Argument	Description
[in]	n	n (integer) is the dimension of the matrices A and B
[in]	A	A is a complex*16 array of dimension (0:n−1,0:n−1)
[in]	B	B is a complex*16 array of dimension (0:n−1,0:n−1)
[out]	C	C is a complex*16 array of dimension (0:n−1,0:n−1), which is the commutator between matrices A and B

Implementation

```
subroutine COMMC(n,A,B,C)
implicit none
integer::n
complex*16::A(0:n-1,0:n-1),B(0:n-1,0:n-1),C(0:n-1,0:n-1)
C=matmul(A,B)-matmul(B,A)
end subroutine
```

Example

In this example we are finding the commutator between two matrices as given below:

$$A = \begin{pmatrix} 1 & 1-i \\ 2+3i & 4 \end{pmatrix}, \quad B = \begin{pmatrix} 1+i & 1+i \\ 3-3i & 4-i \end{pmatrix}. \tag{2.21}$$

```
program example
implicit none
complex*16::A(0:1,0:1),B(0:1,0:1),C(0:1,0:1)
integer::i,j
A(0,0)=dcmplx(1,0)
A(0,1)=dcmplx(1,-1)
A(1,0)=dcmplx(2,3)
A(1,1)=dcmplx(4,0)
B(0,0)=dcmplx(1,1)
B(0,1)=dcmplx(1,1)
B(1,0)=dcmplx(3,-3)
B(1,1)=dcmplx(4,-1)
call COMMC(2,A,B,C)
do i=0,1
   do j=0,1
      write(111,*) C(i,j)
```

```
      end do
end do
end program
```

The file "fort.111" will contain the commutator between matrices A and B.

```
        (1.0000000000000000,-11.000000000000000)
        (-2.0000000000000000,-8.0000000000000000)
        (-3.0000000000000000,-14.000000000000000)
        (-1.0000000000000000,11.000000000000000)
```

2.9   Anticommutator between two matrices

The anticommutator C between two matrices A and B is defined as below:

$$C = \{A, B\} = AB + BA, \tag{2.22}$$

if $\{A, B\} = 0 \Rightarrow AB = -BA$, then matrices A and B are said to anticommute with each other. Anticommutators occur in the areas of second quantization involving Fermionic and Bosonic operators [15]. The flowcharts of finding anticommutator of real and complex matrices are the same.

2.9.1   For real matrices

```
subroutine ANTICOMMR(n,A,B,C)
```

Parameters

In/Out	Argument	Description
[in]	n	n (integer) is the dimension of the matrices A and B
[in]	A	A is a real*8 array of dimension $(0{:}n{-}1,0{:}n{-}1)$
[in]	B	B is a real*8 array of dimension $(0{:}n{-}1,0{:}n{-}1)$
[out]	C	C is a real*8 array of dimension $(0{:}n{-}1,0{:}n{-}1)$ which is the anti commutator between matrices A and B

Implementation

```
subroutine ANTICOMMR(n,A,B,C)
implicit none
integer::n
real*8::A(0:n-1,0:n-1),B(0:n-1,0:n-1),C(0:n-1,0:n-1)
C=matmul(A,B)+matmul(B,A)
end subroutine
```

Example

In this example we are finding the anti commutator between two matrices as given below:

$$A = \begin{pmatrix} 1 & 2 \\ 2 & -1 \end{pmatrix}, \quad B = \begin{pmatrix} 1 & 2 \\ 3 & 1 \end{pmatrix}. \qquad (2.23)$$

```
program example
implicit none
real*8::A(0:1,0:1),B(0:1,0:1),C(0:1,0:1)
integer::i
A(0,:)=(/1.0,2.0/)
A(1,:)=(/2.0,-1.0/)
B(0,:)=(/1.0,2.0/)
B(1,:)=(/3.0,1.0/)
call ANTICOMMR(2,A,B,C)
do i=0,1,1
write(111,*) C(i,:)
end do
end program
```

The file "fort.111" will contain the anti commutator between matrices A and B.

12.000000000000000	4.0000000000000000
4.0000000000000000	8.0000000000000000

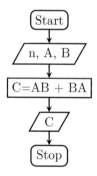

FIGURE 2.13: Flowchart of ANTICOMMR and ANTICOMMC

2.9.2 For complex matrices

```
subroutine ANTICOMMC(n,A,B,C)
```

Parameters

In/Out	Argument	Description
[in]	n	n (integer) is the dimension of the matrices A and B
[in]	A	A is a complex*16 array of dimension (0:n−1,0:n−1)
[in]	B	B is a complex*16 array of dimension (0:n−1,0:n−1)
[out]	C	C is a complex*16 array of dimension (0:n−1,0:n−1) which is the anti commutator between matrices A and B

Implementation

```
subroutine ANTICOMMC(n,A,B,C)
implicit none
integer::n
complex*16::A(0:n-1,0:n-1),B(0:n-1,0:n-1),C(0:n-1,0:n-1)
C=matmul(A,B)+matmul(B,A)
end subroutine
```

Example

In this example we are finding the anti commutator between two matrices as given below:

$$A = \begin{pmatrix} 1 & 1-i \\ 2+3i & 4 \end{pmatrix}, \quad B = \begin{pmatrix} 1+i & 1+i \\ 3-3i & 4-i \end{pmatrix}. \tag{2.24}$$

```
program example
implicit none
complex*16::A(0:1,0:1),B(0:1,0:1),C(0:1,0:1)
integer::i,j
A(0,0)=dcmplx(1,0)
A(0,1)=dcmplx(1,-1)
A(1,0)=dcmplx(2,3)
A(1,1)=dcmplx(4,0)
B(0,0)=dcmplx(1,1)
B(0,1)=dcmplx(1,1)
B(1,0)=dcmplx(3,-3)
B(1,1)=dcmplx(4,-1)
call ANTICOMMC(2,A,B,C)
do i=0,1
   do j=0,1
      write(111,*) C(i,j)
   end do
end do
end program
```

The file "fort.111" will contain the anti commutator between matrices A and B.

```
(1.0000000000000000,1.0000000000000000)
(10.000000000000000,0.0000000000000000)
(25.000000000000000,0.0000000000000000)
(31.000000000000000,-9.0000000000000000)
```

2.10   Binary to decimal conversion

In mathematics, binary numbers are expressed in the *base-2* number system, which only uses '0' and '1' to represent them, which are called bits. Any binary number can be easily converted to a decimal using bit values as follows. Let us have a binary string $[a_1 a_2 a_3 \cdots a_n]_2$ where values a_i's can be either 0 or 1 only. To convert a binary string into the decimal equivalent we have,

$$[a_1 a_2 a_3 \cdots a_n]_2 \equiv a_1 2^{n-1} + a_2 2^{n-2} + \cdots + a_{n-1} 2^1 + a_n 2^0, \qquad (2.25)$$

here the right to left convention is followed. Binary to decimal conversion is a very important operation in quantum mechanics and quantum information theory because the states of the quantum systems (meaning pure states) are represented by a string of zeros and ones. For example, the state of two electrons can be represented as $|00\rangle = |\uparrow\uparrow\rangle$, $|01\rangle = |\uparrow\downarrow\rangle$, $|10\rangle = |\downarrow\uparrow\rangle$ and $|11\rangle = |\downarrow\downarrow\rangle$. When we have to manipulate such states, then the bit representation of such states and their corresponding decimal equivalent becomes handy. For example, the decimal equivalent of the above states can be written as $|00\rangle = |0\rangle$, $|01\rangle = |1\rangle$, $|10\rangle = |2\rangle$ and $|11\rangle = |3\rangle$. Such representations can be easily extended to multi qubit states, for example, a state with four electrons can be in one of the states given by $|1010\rangle = |\downarrow\uparrow\downarrow\uparrow\rangle = |10\rangle$. The main advantage of this decimal representation is that, instead of dealing with a string of numbers which may often be tiring, we deal only with one number. It is remarked here that, this subroutine will be used very frequently in this book.

subroutine BTOD(m,n,s)

Parameters

In/Out	Argument	Description
[in]	m	m is an integer array of dimension (0:s−1) which is the given binary string
[in]	s	s (integer) is the total number of given bits in the string
[out]	n	n (integer) is the decimal equivalent of the given binary string

Implementation

```
subroutine BTOD(m,n,s)
implicit none
```

```
integer*4::s
integer,dimension(0:s-1)::m
integer::n,i,k
n=0
do i=0,s-1
    k=(2**(i))*(m(s-1-i))
n=n+k
end do
end subroutine
```

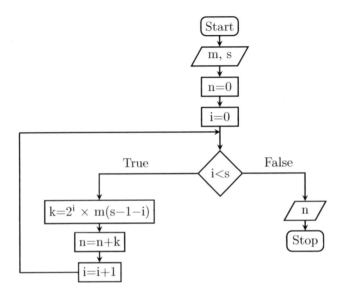

FIGURE 2.14: Flowchart of BTOD

Example

In this example we are converting 1010 into a decimal number.

```
program example
implicit none
integer::m(0:3),n
m=(/1,0,1,0/)
call BTOD(m,n,4)
write(111,*) n
end program
```

The file "fort.111" will contain the decimal equivalent of 1010.

10

2.11 Decimal to binary conversion

To convert an integer number n into a binary string, we adopt the following algorithm,

1. Divide the integer n by 2.
2. Store the remainder from step 1 as the least significant number of the binary string.
3. Divide the quotient (which we got from previous step) by 2, assign the remainder as the second least significant number.
4. Repeat the process till we get the quotient to be zero.

The final remainder will be the most significant digit of the binary string.

This is a crucial subroutine as, if we want to reconvert the state of the quantum system back into its naive form involving bit strings, this routine is very useful and not to mention, it occurs frequently in this book.

```
subroutine DTOB(m,tt,s)
```

Parameters

In/Out	Argument	Description
[in]	m	m (integer) is an input number (decimal)
[in]	s	s (integer) is the total number of bits in the binary string required
[out]	tt	tt is an integer array of dimension (0:s−1) which stores the binary equivalent string of the given integer

Implementation

```
subroutine DTOB(m,tt,s)
implicit none
integer::s
integer,dimension(0:s-1)::tt
integer::m,k,a2
tt=0
a2=m
do k = 0,s-1,1
    tt(s-k-1) = mod(a2,2)
    a2 = a2/2
    if (a2== 0) then
        exit
    end if
end do
end subroutine
```

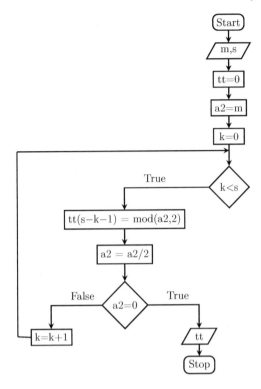

FIGURE 2.15: Flowchart of DTOB

Example

In this example we are converting the integer 10 into a binary number.

```
program example
implicit none
integer::m(0:3),n
n=10
call DTOB(n,m,4)
write(111,*) m
end program
```

The file "fort.111" will contain the binary equivalent of 10.

1	0	1	0

2.12   Bit test

Bit test is a very useful routine in testing whether a particular bit in the given binary string is '0' or '1'. For example, if we have a binary string given by 1010 and we label the place value using the index as subscripts like $1_0 0_1 1_2 0_3$, then, the following routine will give you the bit corresponding to the particular location or index. Suppose we test the bit in location 2 (for labeling we are considering zero-based indexing), it will give us the answer as 1.

subroutine DTOBONEBIT(m,ia,i0,s)

Parameters

In/Out	Argument	Description
[in]	m	m (integer) is an input number (decimal)
[in]	s	s (integer) is the total number of bits in the binary string required to be tested
[in]	i0	i0 (integer) is the position or label of the bit to be tested
[out]	ia	ia (integer) is the value of the bit corresponding to i0 after testing it

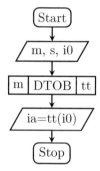

FIGURE 2.16: Flowchart of DTOBONEBIT

Implementation

```
subroutine c(m,ia,i0,s)
implicit none
integer*4::s
integer,dimension(0:s-1)::tt
integer::m,ia,i0
call DTOB(m,tt,s)
ia=tt(i0)
end subroutine
```

Example

In this example we are finding the bit value corresponding to the label 1 of the binary string corresponding to the decimal number equal to 10.

```
program example
implicit none
```

```
integer::n,ia,i0
n=10
i0=1
call DTOBONEBIT(n,ia,i0,4)
write(111,*) ia
end program
```

The file "fort.111" will contain the bit in its second position corresponding to label 1.

0

2.13   Left shift

This algorithm is used in shifting the string of binary digits to the left by k place values. Let us have a shift operator S_L as given below:

$$S_L[a_1a_2a_3\cdots a_n]_2 = [a_2a_3a_4\cdots a_na_0]_2, \qquad (2.26)$$

here a_i's take the value either 0 or 1 in the binary representation. For shifting k times we have to operate S_L totally k times in succession. This routine is useful for studying quantum states which are shift symmetric and it can be put to other generic uses depending on the problem under consideration.

```
subroutine LEFTSHIFT(a,s,k)
```

Parameters

In/Out	Argument	Description
[in/out]	a	a is an integer array of dimension (0:s−1) which is the given binary string, on exit it will be overwritten by the shifted string
[in]	s	s (integer) is the total number of bits in the binary string required
[in]	k	k (integer) is the total number of left shifts required

Implementation

```
subroutine LEFTSHIFT(a,s,k)
implicit none
integer::s,k
integer,dimension(0:s-1)::a,b
integer::i,j
do j=0,k-1
    do i=0,s-2,1
        b(i)=a(i+1)
```

```
    end do
    b(s-1)=a(0)
    a=b
    b=0
end do
end subroutine
```

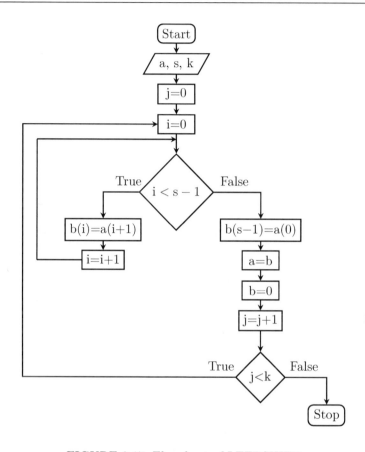

FIGURE 2.17: Flowchart of LEFTSHIFT

Example

In this example we are shifting the binary string 0110 to the left by two places.

```
program example
implicit none
integer::m(0:3)
m=(/0,1,1,0/)
call LEFTSHIFT(m,4,2)
write(111,*) m
end program
```

The file "fort.111" will contain the left shifted binary string by two places.

1	0	0	1

2.14   Right shift

This algorithm is used in shifting the string of binary digits to the right by k place values. Let us have a shift operator S_R as given below:

$$S_R[a_1a_2a_3\cdots a_n]_2 = [a_n a_0 a_1 a_2 a_3 \cdots a_{n-1}]_2, \tag{2.27}$$

here a_i's take the value either 0 or 1 in the binary representation. For shifting k times we have to operate S_R totally k times in succession.

```
subroutine RIGHTSHIFT(a,s,k)
```

Parameters

In/Out	Argument	Description
[in/out]	a	a is an integer array of dimension (0:s−1) which is the given binary string, on exit it will be overwritten by the shifted string
[in]	s	s (integer) is the total number of bits in the binary string required
[in]	k	k (integer) is the total number of right shifts required

Implementation

```
subroutine RIGHTSHIFT(a,s,k)
implicit none
integer::s,k
integer,dimension(0:s-1)::a,b
integer::i,j
do j=0,k-1
    do i=0,s-2,1
        b(i+1)=a(i)
    end do
    b(0)=a(s-1)
    a=b
    b=0
end do
end subroutine
```

Example

In this example we are shifting the binary string 0110 to the right by two places.

```
program example
implicit none
integer::m(0:3)
m=(/0,1,1,0/)
```

```
call RIGHTSHIFT(m,4,2)
write(111,*) m
end program
```

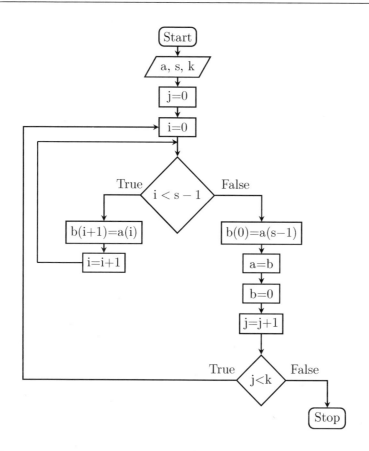

FIGURE 2.18: Flowchart of RIGHTSHIFT

The file "fort.111" will contain the right shifted binary string by two places.

1	0	0	1

2.15 Swapping or permutation operator

This algorithm is used in swapping the position of ith and jth bit in the string of binary which is representing the quantum state. This swapping is done by the operator P_{ij} as shown below:

$$P_{ij}[a_1a_2\cdots a_i\cdots a_j\cdots a_n]_2 = [a_1a_2\cdots a_j\cdots a_i\cdots a_n]_2, \qquad (2.28)$$

here a_i's take the value either 0 or 1 in the binary representation.

```
subroutine SWAP(s,i,j,a)
```

Parameters

In/Out	Argument	Description
[in]	s	s (integer) is the total number of bits in the binary string required.
[in]	i, j	i and j (integer) are the labels of the corresponding bits which are involved in the swap operation
[in/out]	a	a is an integer array of dimension (0:s−1) which is the given binary string, on exit it will be overwritten by the swapped string

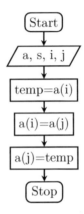

FIGURE 2.19: Flowchart of SWAP

Implementation

```
subroutine SWAP(s,i,j,a)
implicit none
integer::s,i,j,temp
integer,dimension(0:s-1)::a
temp=a(i)
a(i)=a(j)
a(j)=temp
end subroutine
```

Example

In this example we are swapping the first and second entries of the binary string 1010.

```
program example
implicit none
integer::m(0:3)
m=(/1,0,1,0/)
call SWAP(4,0,1,m)
```

```
write(111,*) m
end program
```

The file "fort.111" will contain the swapped binary string.

0	1	1	0

3

Numerical Linear Algebra and Matrix Operations

Having done with the basic recipes involving vector and matrix elements in the previous chapter, here in this chatper we will give the recipes for some major linear algebra operations [13, 16, 17]. These operations are very important from the point of view of many quantum information tasks and also required in the solution of major problems in quantum mechanics and elsewhere. To be more specific, this chapter broadly deals with matrix decompositions, calculation of different matrix norms, functions of matrices and orthogonalization procedures [12, 18]. Linear algebra is an ocean, but we have just sailed in that part of it which is the most important for quantum information theory. The codes given are optimized and directly usable along with their dependencies. In the forthcoming sections we will use the following matrices for the purpose of illustration of how the recipes work. for a real symmetric matrix,

$$A_1 = \begin{pmatrix} 1 & -2 & 3 \\ -2 & 3 & 4 \\ 3 & 4 & 5 \end{pmatrix}. \tag{3.1}$$

For a complex Hermitian matrix,

$$A_2 = \begin{pmatrix} 2 & 1-3i \\ 1+3i & 4 \end{pmatrix}. \tag{3.2}$$

For a real non-symmetric matrix,

$$A_3 = \begin{pmatrix} 1 & -2 & 3 \\ 2 & 3 & 5 \\ -4 & 4 & 5 \end{pmatrix}. \tag{3.3}$$

For a complex non-symmetric matrix,

$$A_4 = \begin{pmatrix} 2 & 1-3i \\ 2+i & 4 \end{pmatrix}. \tag{3.4}$$

3.1 Inverse of a matrix

Finding the inverse of a matrix is a very important operation in matrix algebra. This originated from the efforts to solve a system of n simultaneous linear equations for n unknowns.

Let us now consider $n \times n$ matrix A is called invertible if we can find a matrix A^{-1} such that,

$$AA^{-1} = A^{-1}A = \mathbb{I}. \tag{3.5}$$

A^{-1} can be found using the relation given below,

$$A^{-1} = \frac{adj(A)}{det(A)}, \tag{3.6}$$

where $adj(A)$ is the transpose of its co-factor matrix. A^{-1} also can be found using the LU decomposition which is a decomposition of a given square matrix A into a product of lower and upper triangular matrices which are denoted by L and U, respectively, we can write,

$$A = LU \implies A^{-1} = (LU)^{-1} = U^{-1}L^{-1}. \tag{3.7}$$

You may wonder, what is the advantage of finding the inverse of a matrix using the LU decomposition? The answer is simply the fact that, the computational time required to find the inverse using LU decomposition is less than the usual methods such as direct Gauss elimination, etc., specifically for large dimension matrices.

3.1.1 For a real matrix

---
subroutine MATINVMR(n,A,B)

Parameters

In/Out	Argument	Description
[in]	n	n (integer) is the dimension of the matrices A and B
[in]	A	A is a real*8 array of dimension (0:n−1,0:n−1), which is the given matrix
[out]	B	B is a real*8 array of dimension (0:n−1,0:n−1), which is the inverse of the matrix A

Implementation

```
subroutine MATINVMR(n,A,B)
implicit none
integer::n
real*8,dimension(0:n-1,0:n-1)::A,B
integer::LWORK,INFO,i
integer,dimension(0:n-1)::IPIV
double precision::WORK(0:n-1)
B=A
call DGETRF(n, n, B, n, IPIV, INFO)
call DGETRI(n, B, n, IPIV, WORK, n, INFO)
end subroutine
```

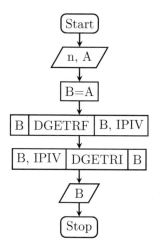

FIGURE 3.1: Flowchart of MATINVMR

Example

In this example we will find the inverse of the matrix given below:

$$A = \begin{pmatrix} 1 & 2 & 3 \\ 3 & -2 & -5 \\ 7 & -2 & -3 \end{pmatrix}. \tag{3.8}$$

```
program example
implicit none
real*8::A(0:2,0:2),B(0:2,0:2)
integer::i
A(0,:)=(/1.0,2.0,3.0/)
A(1,:)=(/3.0,-2.0,-5.0/)
A(2,:)=(/7.0,-2.0,-3.0/)
call MATINVMR(3,A,B)
do i=0,2,1
write(111,*)B(i,:)
end do
end program
```

The file "fort.111" contains the inverse of the matrix A.

0.12500000000000000	0.0000000000000000	0.12500000000000000
0.81250000000000000	0.75000000000000000	-0.43749999999999994
-0.25000000000000000	-0.50000000000000000	0.25000000000000000

3.1.2 For a complex matrix

```
subroutine MATINVMC(n,A,B)
```

Parameters

In/Out	Argument	Description
[in]	n	n (integer) is the dimension of the matrices A and B
[in]	A	A is a complex*16 array of dimension (0:n−1,0:n−1), which is the given matrix
[out]	B	B is a complex*16 array of dimension (0:n−1,0:n−1), which is the inverse of matrix A

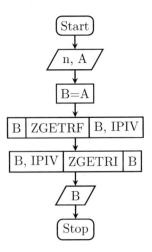

FIGURE 3.2: Flowchart of MATINVMC

Implementation

```
subroutine MATINVMC(n,A,B)
implicit none
integer::n
complex*16,dimension(0:n-1,0:n-1)::A,B
integer::LWORK,INFO
integer,dimension(0:n-1)::IPIV
complex*16::WORK(0:60*n-1)
B=A
call ZGETRF(n, n, B, n, IPIV, INFO)
call ZGETRI(n, B, n, IPIV, WORK, 60*n, INFO)
end subroutine
```

Example

In this example we will find the inverse of the matrix given below:

$$A = \begin{pmatrix} 1+i & 1+i \\ 3-3i & 4-i \end{pmatrix}. \tag{3.9}$$

```
program example
implicit none
complex*16::B(0:1,0:1),A(0:1,0:1)
integer::i,j
A(0,0)=dcmplx(1,1)
A(0,1)=dcmplx(1,1)
A(1,0)=dcmplx(3,-3)
A(1,1)=dcmplx(4,-1)
call MATINVMC(2,A,B)
do i=0,1,1
    do j=0,1,1
        write(111,*)B(i,j)
    end do
end do
end program
```

The file "fort.111" contains the inverse of the matrix A.

```
(-0.70000000000000007,-1.099999999999999)
(-0.1999999999999993,0.40000000000000002)
 (1.2000000000000002,0.59999999999999987)
 (0.1999999999999996,-0.40000000000000002)
```

3.2 Lanczos tridiagonalization

Lanczos tridiagonaliztion [19] is the process which transforms any Hermitian matrix A into a tridiagonal form T. By tridiagonal form we mean, there are elements along the main diagonal and the two adjacent diagonals along the either side of the main diagonal, however all the other elements are zero. Tridiagonalization is basically a basis change which is accomplished using a unitary transformation as shown below. The unitary matrix U which transform matrix $A_{n\times n}$ into tridiagonal matrix $T_{n\times n}$ as shown below,

$$U A U^\dagger = T = \begin{pmatrix} a_1 & b_2 & & & & & 0 \\ b_2 & a_2 & b_3 & & & & \\ & b_3 & a_3 & \ddots & & & \\ & & \ddots & \ddots & b_{n-1} & & \\ & & & b_{n-1} & a_{n-1} & b_n & \\ 0 & & & & b_n & a_n \end{pmatrix}. \tag{3.10}$$

In case of A being real symmetric matrix, U will be become an orthogonal matrix. Tridiagonalization is important in various branches of science, the tridiagonal matrix T is a sparse matrix, meaning it has many zeros. Due to this fact we have to deal with very less number of non-zero elements and it has great advantages when we find eigenvalues and eigenvectors of Hermitian matrices and offers tangible reduction in computational time.

3.2.1 For a real symmetric matrix

```
subroutine LANCZOSSR(n,A,D,E)
```

Parameters

In/Out	Argument	Description
[in]	n	n (integer) is the dimension of the matrix A
[in/out]	A	A is a real*8 array of dimension $(0{:}n{-}1,0{:}n{-}1)$ on exit it will give an orthogonal matrix which tridiagonalizes A
[out]	D	D is a real*8 array of dimension $(0{:}n{-}1)$ with only the diagonal elements of the tridiagonal matrix
[out]	E	E is a real*8 array of dimension $(0{:}n{-}2)$ which contains both upper and lower off-diagonal elements which are equal

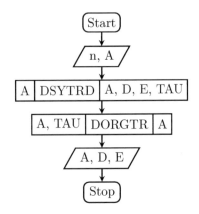

FIGURE 3.3: Flowchart of LANCZOSSR

Implementation

```
subroutine LANCZOSSR(n,A,D,E)
implicit none
integer::n,INFO
real*8::A(0:n-1,0:n-1),D(0:n-1),E(0:n-2),TAU(0:n-1),WORK(0:n-1)
call DSYTRD('U', n, A, n, D, E, TAU, WORK,n, INFO)
call DORGTR('U', n, A, n, TAU, WORK, n, INFO)
```

end subroutine

Example

In this example we are doing the trdiagonalization of the matrix given below:

$$A = \begin{pmatrix} 1 & -2 & 3 & 4 \\ -2 & 3 & 4 & -5 \\ 3 & 4 & 5 & 6 \\ 4 & -5 & 6 & -7 \end{pmatrix}. \tag{3.11}$$

```
program example
implicit none
real*8::A(0:3,0:3),D(0:3),E(0:2)
integer::i
A(0,:)=(/1,-2,3,4/)
A(1,:)=(/-2,3,4,-5/)
A(2,:)=(/3,4,5,6/)
A(3,:)=(/4,-5,6,-7/)
call LANCZOSSR(4,A,D,E)
write(111,*) 'Diagonal elements:'
do i=0,3,1
   write(111,*)D(i)
end do
write(111,*) 'Off-diagonal elements:'
do i=0,2,1
   write(111,*)E(i)
end do
end program
```

The file "fort.111" contains the tridiagonal entries.

```
Diagonal elements:
  4.9043151969981240
  0.78399649131356064
  3.3116883116883127
 -7.0000000000000000
Off-diagonal elements:
  4.4780306066991677
  2.9377071201244038
 -8.7749643873921208
```

3.2.2   For a complex Hermitian matrix

subroutine LANCZOSHC(n,A,D,E)

Parameters

In/Out	Argument	Description
[in]	n	n (integer) is the dimension of the matrix A
[in/out]	A	A is a complex*16 array of dimension (0:n−1,0:n−1) on exit it will give a unitary matrix which tridiagonalizes A
[out]	D	D is a real*8 array of dimension (0:n−1) which contains only the diagonal elements of the tridiagonal matrix
[out]	E	E is a real*8 array of dimension (0:n−2) which contains both upper and lower off-diagonal elements which are equal

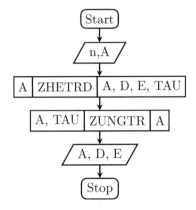

FIGURE 3.4: Flowchart of LANCZOSHC

Implementation

```
subroutine LANCZOSHC(n,A,D,E)
implicit none
integer::n,INFO
complex*16::A(0:n-1,0:n-1)
complex*16::TAU(0:n-1),WORK(0:n-1)
real*8::D(0:n-1),E(0:n-2)
call ZHETRD('U', n, A, n, D, E, TAU, WORK, n, INFO)
call ZUNGTR('U', n, A, n, TAU, WORK, n, INFO)
end subroutine
```

Example

In this example we are doing the trdiagonalization of the matrix given below:

$$A = \begin{pmatrix} 2 & 2+i & 5+3i \\ 2-i & 4 & 2-3i \\ 5-3i & 2+3i & 4 \end{pmatrix}. \tag{3.12}$$

```
program example
implicit none
real*8::D(0:3),E(0:2)
integer::i
complex*16::A(0:3,0:3)
A(0,0)=dcmplx(2,0)
A(0,1)=dcmplx(2,1)
A(0,2)=dcmplx(5,3)
A(1,0)=dcmplx(2,-1)
A(1,1)=dcmplx(4,0)
A(1,2)=dcmplx(2,-3)
A(2,0)=dcmplx(5,-3)
A(2,1)=dcmplx(2,3)
A(2,2)=dcmplx(4,0)
call LANCZOSHC(3,A,D,E)
write(111,*) 'Diagonal elements:'
do i=0,2,1
    write(111,*)D(i)
end do
write(111,*) 'Off-diagonal elements:'
do i=0,1,1
    write(111,*)E(i)
end do
end program
```

The file "fort.111" contains tridiagonal entries

```
Diagonal elements:
  1.9999999999999993
  1.9999999999999996
  5.0000000000000000
Off-diagonal elements:
  5.3353123211448278E-016
 -3.6055512754639896
```

Note that even though matrix A is having complex entries, the tridiagonal form will have only real entries.

3.3   QR decomposition

Any complex rectangular $n \times k$ matrix A can be decomposed as follows,

$$A = QR, \tag{3.13}$$

where $Q_{n \times k}$ is a unitary matrix and $R_{k \times k}$ is an upper triangular matrix [20]. Explicitly writing we have,

$$A = \begin{pmatrix} q_1 & q_2 & \cdots & q_k \end{pmatrix} \begin{pmatrix} R_{11} & R_{12} & \cdots & R_{1k} \\ 0 & R_{22} & \cdots & R_{2k} \\ \vdots & \vdots & \ddots & \vdots \\ 0 & 0 & \cdots & R_{kk} \end{pmatrix}. \tag{3.14}$$

In the above decomposition it is important to note that, the matrix $Q = \begin{pmatrix} q_1 & q_2 & \cdots & q_k \end{pmatrix}$ are orthonormal column vectors such that $q_i^\dagger q_j = 0$ for $i \neq j$ and $||q_i|| = 1$ (unit norm). The diagonal elements of R are such that $R_{ii} \neq 0$. Since matrix Q is of dimension $n \times k$ it has k orthonormal column vectors contained in it. QR decomposition finds many uses in areas like solution of linear equations, to generate orthonormal vectors, etc.

3.3.1   For a real matrix

subroutine QRMR(n,k,A,R)

Parameters

In/Out	Argument	Description
[in]	n	n (integer) is the number of rows in matrix A
[in]	k	k (integer) is the number of columns in matrix A
[in/out]	A	A is a real*8 array of dimension (0:n−1,0:k−1), on exit, gives an orthogonal matrix Q
[out]	R	R is a real*8 array of dimension (0:k−1,0:k−1) which is an upper triangular matrix

Implementation

```
subroutine QRMR(n,k,A,R)
implicit none
integer::n,k,INFO,i,j,temp
real*8::A(0:n-1,0:k-1),R(0:k-1,0:k-1)
real*8::WORK(0:k-1)
real*8,allocatable,dimension(:)::TAU
R=0.0d0
temp=min(n,k)
allocate(TAU(0:temp-1))
call DGEQRF(n, k, A, n, TAU, WORK, k, INFO)
do i=0,k-1
    do j=i,k-1
```

```
        R(i,j)=A(i,j)
    end do
end do
call DORGQR(n, k, k, A, n, TAU, WORK, n, INFO)
deallocate(TAU)
end subroutine
```

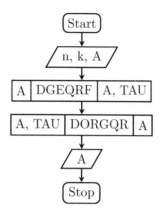

FIGURE 3.5: Flowchart of QRMR

Example

In this example we are doing the QR decomposition of the matrix A_1.

```
program example
implicit none
real*8::A(0:2,0:2),R(0:2,0:2)
integer::i
A(0,:)=(/1,-2,3/)
A(1,:)=(/-2,3,4/)
A(2,:)=(/3,4,5/)
call QRMR(3,3,A,R)
write(111,*)'matrix Q'
do i=0,2,1
    write(111,*)real(A(i,:))
end do
write(111,*)'matrix R'
do i=0,2,1
    write(111,*)real(R(i,:))
end do
end program
```

The file "fort.111" contains matrices Q and R.

```
  matrix Q
-0.267261237      0.433065563     -0.860828459
 0.534522474     -0.676664948     -0.506369710
-0.801783741     -0.595465183     -5.06369695E-02
  matrix R
```

```
    -3.74165750        -1.06904495        -2.67261243
     0.00000000        -5.27798653        -4.38478899
     0.00000000         0.00000000        -4.86114883
```

3.3.2   For a complex matrix

```
subroutine QRMC(n,k,A,R)
```

Parameters

In/Out	Argument	Description
[in]	n	n (integer) is the number of rows in matrix A
[in]	k	k (integer) is the number of columns in matrix A
[in/out]	A	A is a complex*16 array of dimension (0:n−1,0:k−1) on exit, gives the unitary matrix
[out]	R	R is a complex*16 array of dimension (0:k−1,0:k−1) which is an upper triangular matrix

Flowchart

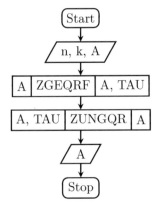

FIGURE 3.6: Flowchart of QRMC

Implementation

```
subroutine QRMC(n,k,A,R)
implicit none
integer::n,k,INFO,i,j,temp
complex*16::A(0:n-1,0:k-1),R(0:k-1,0:k-1)
complex*16::WORK(0:k-1)
complex*16,allocatable,dimension(:)::TAU
```

```
R=0.0d0
temp=min(n,k)
allocate(TAU(0:temp-1))
call ZGEQRF(n, k, A, n, TAU, WORK, k, INFO)
do i=0,k-1
   do j=i,k-1
      R(i,j)=A(i,j)
   end do
end do
call ZUNGQR(n, k, k, A, n, TAU, WORK, k, INFO)
deallocate(TAU)
end subroutine
```

Example

In this example we are doing the QR decomposition of the matrix A_2.

```
program example
implicit none
complex*16::A(0:1,0:1),R(0:1,0:1)
integer::i,j
A(0,0)=dcmplx(2)
A(0,1)=dcmplx(1,-3)
A(1,0)=dcmplx(1,3)
A(1,1)=dcmplx(4,0)
call QRMC(2,2,A,R)
write(111,*)'matrix Q'
do i=0,1,1
   do j=0,1,1
      write(111,*)A(i,j)
   end do
end do
write(111,*)'matrix R'
do i=0,1,1
   do j=0,1,1
      write(111,*)R(i,j)
   end do
end do
end program
```

The file "fort.111" contains matrices Q and R.

```
 matrix Q
            (-0.53452248382484857,0.0000000000000000)
            (0.26726124191242473,-0.80178372573727297)
            (-0.26726124191242440,-0.80178372573727308)
     (-0.53452248382484879,-2.21118388796120877E-016)
 matrix R
            (-3.7416573867739413,0.0000000000000000)
            (-1.6035674514745462,4.8107023544236371)
            (0.0000000000000000,0.0000000000000000)
            (0.53452248382484768,0.0000000000000000)
```

3.4　Function of a matrix

Given any square matrix M of dimension $n \times n$, we can ask the question, what will be the logarithm of the matrix? [21] or what will be the power of the matrix? It is known that any real continuous function $f(x)$ or a complex analytic function $f(z)$ has a Taylor series expansion around a point a given by,

$$f(x) = \sum_{n=0}^{\infty} \frac{f^{(n)}(a)}{n!}(x-a)^n, \tag{3.15}$$

if the expansion is done around $a = 0$, it is called a Maclaurin expansion which has the form,

$$f(x) = \sum_{n=0}^{\infty} \frac{f^{(n)}(0)}{n!}x^n. \tag{3.16}$$

Just like any function above, any function of matrix M also can be expanded using the Maclaurin expansion as follows,

$$f(M) = \mathbb{I}f(0) + \frac{f^1(0)}{1!}M + \frac{f^2(0)}{2!}M^2 + \frac{f^3(0)}{3!}M^3 + \cdots = \sum_{n=0}^{\infty}\frac{f^n(0)}{n!}M^n. \tag{3.17}$$

If the matrix is diagonalizable, the above equation reduces to a matrix form given as below,

$$f(M) = Pf(D)P^{-1}, \tag{3.18}$$

where P is a matrix whose columns are made the of normalized eigenvectors of M and D is a diagonal matrix which contains the eigenvalues of M as diagonal elements. If M is a normal matrix Eq. [3.17] is reduced to,

$$f(M) = Uf(D)U^{\dagger}, \tag{3.19}$$

where U is a unitary matrix. As we know almost all the problems of physics and specially quantum mechanics and quantum information theory involve only Hermitian matrices, they are 'compulsorily' diagonalizable, because every Hermitian matrix is a normal matrix which can be diagonalized by a unitary transform. Thus Eq. [3.19] can be used to find the function of matrices. We require the exponential of a matrix when we study the time evolution of quantum systems, in finding quantum entropies we require the logarithm of a matrix and in many other diverse areas other functions of matrices like power and trigonometric functions are used. Note that for matrices which are not diagonalizable, these techniques can not be used and in such cases the power series with proper truncation is used based on the accuracy of the results we require.

3.4.1　For a real symmetric matrix

```
subroutine FUNSR(A,n,F,B)
```

Parameters

In/Out	Argument	Description
[in/out]	A	A is a real*8 array of dimension (0:n−1,0:n−1), on exit, the columns of A will be orthonormal eigenvectors of A

In/Out	Argument	Description
[in]	n	n (integer) is the dimension of matrices A and B
[in]	F	F (character*3) is the type of function of matrix which we are interested in; if F='sin', it will compute sin(A) if F='cos', it will compute cos(A) if F='tan', it will compute tan(A) if F='exp', it will compute exp(A) if F='log', it will compute log(A)
[out]	B	B is a real*8 array of dimension $(0{:}n{-}1,0{:}n{-}1)$ which will be the required function of the matrix

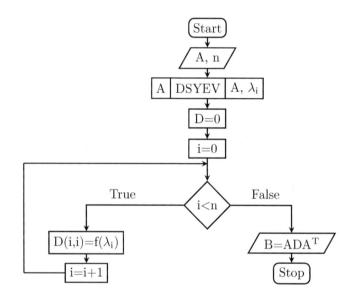

FIGURE 3.7: Flowchart of FUNSR

Implementation

```
subroutine FUNSR(A,n,F,B)
implicit none
real*8::A(0:n-1,0:n-1),B(0:n-1,0:n-1),D(0:n-1,0:n-1)
integer::n,i
character*3::JOBZ, UPLO ,F
integer::INFO
real*8::W(0:N-1),WORK(0:3*N-2),pi
JOBZ='V'
UPLO='U'
pi=4.0d0*atan(1.0d0)
call DSYEV(JOBZ, UPLO, N, A, N, W, WORK, 3*N-1, INFO)
D=0.0d0
```

```
if (F=='exp') then
    do i=0,n-1
        D(i,i)=exp(W(i))
    end do
end if

if (F=='log') then
    do i=0,n-1
        if (W(i) .le. 0.0000000000001) then
            print*, 'Error occured'
            print*, 'Logarithm of zero and &
                    & negative numbers are not defined'
            exit
        end if
        D(i,i)=log(W(i))
    end do
end if

if (F=='sin') then
    do i=0,n-1
        D(i,i)=sin(W(i))
    end do
end if

if (F=='cos') then
    do i=0,n-1
        D(i,i)=cos(W(i))
    end do
end if

if (F=='tan') then
    do i=0,n-1
        D(i,i)=tan(W(i))
    end do
end if
B=matmul(matmul(A,D),transpose(A))
end subroutine
```

Example

In this example we are finding the exponential of the matrix A_1.

```
program example
implicit none
real*8::A(0:2,0:2),B(0:2,0:2)
integer::i
A(0,:)=(/1,-2,3/)
A(1,:)=(/-2,3,4/)
A(2,:)=(/3,4,5/)
call FUNSR(A,3,'exp',B)
do i=0,2,1
```

```
    write(111,*)B(i,:)
end do
end program
```

The file "fort.111" contains the exponential of the matrix A_1.

168.28088998963304	404.54130674754020	646.45726610378790
404.54130674754020	1251.9967893843975	1854.1144922332526
646.45726610378790	1854.1144922332528	2805.6886486135691

3.4.2 For a complex Hermitian matrix

```
subroutine FUNHC(A,n,F,B)
```

Parameters

In/Out	Argument	Description
[in/out]	A	A is a complex*16 array of dimension (0:n−1,0:n−1), on exit, the columns of A will be orthonormal eigenvectors of A
[in]	n	n (integer) is the dimension of matrices A and B
[in]	F	F (character*3) is the type of function of matrix which we are interested in; if F='sin', it will compute sin(A) if F='cos', it will compute cos(A) if F='tan', it will compute tan(A) if F='exp', it will compute exp(A) if F='log', it will compute log(A)
[out]	B	B is a complex*16 array of dimension (0:n−1,0:n−1) which will be the required function of the matrix

Implementation

```
subroutine FUNHC(A,n,F,B)
implicit none
complex*16::A(0:n-1,0:n-1),B(0:n-1,0:n-1),D(0:n-1,0:n-1)
integer::n,i
character*3::JOBZ, UPLO,F
integer::INFO
real*8::W(0:N-1),RWORK(0:3*N-3)
complex*16::WORK(0:2*N-2)
JOBZ='V'
UPLO='U'
call ZHEEV(JOBZ, UPLO, N, A, N, W, WORK, 2*N-1, RWORK,INFO)
D=0.0d0
```

```fortran
if (F=='exp') then
    do i=0,n-1
        D(i,i)=exp(W(i))
    end do
end if

if (F=='log') then
    do i=0,n-1
        if (W(i) .le. 0.0000000000001) then
            print*, 'Error occured'
            print*, 'Logarithm of zero and &
                    & negative numbers are not defined'
            exit
        end if
        D(i,i)=log(W(i))
    end do
end if

if (F=='sin') then
    do i=0,n-1
        D(i,i)=sin(W(i))
    end do
end if

if (F=='cos') then
    do i=0,n-1
        D(i,i)=cos(W(i))
    end do
end if

if (F=='tan') then
    do i=0,n-1
        D(i,i)=tan(W(i))
    end do
end if
B=matmul(matmul(A,D),conjg(transpose(A)))
end subroutine
```

Example

In this example we are finding the logarithm of the matrix A_2.

```fortran
program example
implicit none
complex*16::A(0:1,0:1),B(0:1,0:1)
integer::i,j
A(0,0)=dcmplx(2,0)
A(0,1)=dcmplx(1,-3)
A(1,0)=dcmplx(1,3)
A(1,1)=dcmplx(4,0)
call FUNHC(A,2,'log',B)
```

```
do i=0,1,1
    do j=0,1,1
        write(111,*)B(i,j)
    end do
end do
end program
```

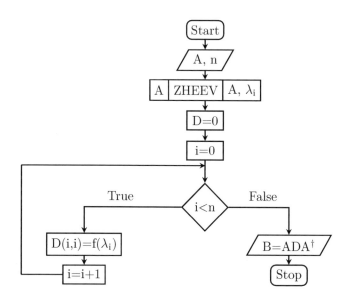

FIGURE 3.8: Flowchart of FUNHC

Since one of the eigenvalue is negative, we will get the following on the terminal.

```
Error occured
 Logarithm of zero and negative numbers are not defined
```

The file "fort.111" contains garbage values.

3.4.3 For a real non-symmetric matrix

```
subroutine FUNMR(A,n,F,E)
```

Parameters

In/Out	Argument	Description
[in]	A	A is a real*8 array of dimension $(0{:}n{-}1,0{:}n{-}1)$
[in]	n	n (integer) is the dimension of matrices A and E

In/Out	Argument	Description
[in]	F	F (character*3) is the type of function of matrix which we are interested in; if F='sin', it will compute sin(A) if F='cos', it will compute cos(A) if F='tan', it will compute tan(A) if F='exp', it will compute exp(A) if F='log', it will compute log(A)
[out]	E	E is a complex*16 array of dimension $(0:n-1,0:n-1)$ which will be the required function of the matrix

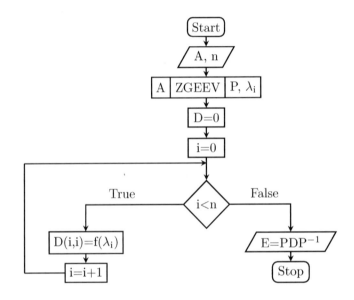

FIGURE 3.9: Flowchart of FUNMR

Implementation

```
subroutine FUNMR(A,n,F,E)
implicit none
integer::n,i,j
complex*16::B(0:n-1,0:n-1),D(0:n-1,0:n-1)
complex*16::E(0:n-1,0:n-1),C(0:n-1,0:n-1)
character*3::JOBVL, JOBVR ,F
integer::INFO
complex*16::W(0:N-1),VL(0:N-1,0:N-1),VR(0:N-1,0:N-1)
complex*16::WORK(0:2*N-1)
real*8::RWORK(0:2*N-1),A(0:N-1,0:N-1)
JOBVL='N'
JOBVR='V'
do i=0,n-1
```

```fortran
      do j=0,n-1
          C(i,j)=dcmplx(A(i,j),0)
      end do
end do
call ZGEEV(JOBVL, JOBVR, N, C, N, W, VL, N, VR, N, &
           WORK, 2*N, RWORK, INFO)
D=0.0d0
if (F=='exp') then
    do i=0,n-1
        D(i,i)=exp(W(i))
    end do
end if

if (F=='log') then
    do i=0,n-1
        if (abs(W(i)) .le. 0.0000000000001) then
            print*, 'Error occured'
            print*, 'Logarithm of zero and &
                 & negative number are not defined'
            exit
        end if
        D(i,i)=log(W(i))
    end do
end if

if (F=='sin') then
    do i=0,n-1
        D(i,i)=sin(W(i))
    end do
end if

if (F=='cos') then
    do i=0,n-1
        D(i,i)=cos(W(i))
    end do
end if

if (F=='tan') then
    do i=0,n-1
        D(i,i)=tan(W(i))
    end do
end if
call MATINVMC(n,VR,B)
E=matmul(matmul(VR,D),B)
end subroutine
```

Example

In this example we are finding the logarithm of the matrix A_3.

```fortran
program example
```

```
implicit none
real*8::A(0:2,0:2)
complex*16::B(0:2,0:2)
integer::i,j
A(0,:)=(/1.0,-2.0,3.0/)
A(1,:)=(/2.0,3.0,5.0/)
A(2,:)=(/-4.0,4.0,5.0/)
call FUNMR(A,3,'log',B)
do i=0,2,1
    do j=0,2,1
        write(111,*)B(i,j)
    end do
end do
end program
```

The file "fort.111" contains the logarithm of the matrix A_3.

```
(1.6443716737846727,1.39693185492382129E-016)
(-0.96712394065694207,-1.84750805486001449E-016)
(1.0181482508907906,1.21941232457043112E-016)
(1.2142883178538704,1.349030491516253334E-016)
(1.2520915398585126,-9.22880285252711313E-017)
(0.68765921126386476,-6.07772736558528740E-017)
(-1.1927547497231026,-3.17758385659964687E-016)
(0.45126161813232052,1.11022302462515654E-016)
(1.8484689147200639,8.63364379924672650E-018)
```

3.4.4   For a complex non-symmetric matrix

```
subroutine FUNMC(A,n,F,E)
```

Parameters

In/Out	Argument	Description
[in/out]	A	A is a complex*16 array of the dimension (0:n−1,0:n−1), on exit A will be overwritten
[in]	n	n (integer) is the dimension of matrices A and E
[in]	F	F (character*3) is the type of function of matrix which we are interested in; if F='sin', it will compute sin(A) if F='cos', it will compute cos(A) if F='tan', it will compute tan(A) if F='exp', it will compute exp(A) if F='log', it will compute log(A)
[out]	E	E is a complex*16 array of dimension (0:n−1,0:n−1) which will be the required function of the matrix

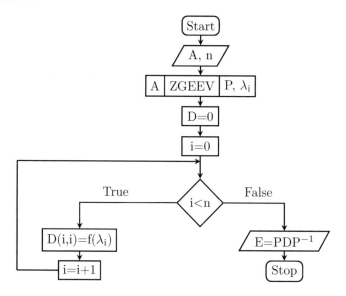

FIGURE 3.10: Flowchart of FUNMC

Implementation

```
subroutine FUNMC(A,n,F,E)
implicit none
integer::n,i
complex*16::B(0:n-1,0:n-1),D(0:n-1,0:n-1)
complex*16::E(0:n-1,0:n-1)
character*3::JOBVL, JOBVR ,F
integer::INFO
complex*16::W(0:N-1),VL(0:N-1,0:N-1),VR(0:N-1,0:N-1)
complex*16::WORK(0:2*N-1),A(0:N-1,0:N-1)
real*8::RWORK(0:2*N-1)
JOBVL='N'
JOBVR='V'
call ZGEEV(JOBVL, JOBVR, N, A, N, W, VL, N, VR, &
          N, WORK, 2*N, RWORK, INFO)
D=0.0d0
if (F=='exp') then
    do i=0,n-1
        D(i,i)=exp(W(i))
    end do
end if

if (F=='log') then
    do i=0,n-1
        if (abs(W(i)) .le. 0.0000000000001) then
            print*, 'Error occured'
            print*, 'Logarithm of zero and &
                & negative number are not defined'
            exit
```

```
         end if
         D(i,i)=log(W(i))
      end do
end if

if (F=='sin') then
   do i=0,n-1
      D(i,i)=sin(W(i))
   end do
end if

if (F=='cos') then
   do i=0,n-1
      D(i,i)=cos(W(i))
   end do
end if

if (F=='tan') then
   do i=0,n-1
      D(i,i)=tan(W(i))
   end do
end if
call MATINVMC(n,VR,B)
 E=matmul(matmul(VR,D),B)
end subroutine
```

Example

In this example we are finding the cosine of the matrix A_4.

```
program example
implicit none
complex*16::A(0:1,0:1),B(0:1,0:1)
integer::i,j
A(0,0)=dcmplx(2,0)
A(0,1)=dcmplx(1,-3)
A(1,0)=dcmplx(2,1)
A(1,1)=dcmplx(4,0)
call FUNMC(A,2,'cos',B)
do i=0,1,1
   do j=0,1,1
      write(111,*)B(i,j)
   end do
end do
end program
```

The file "fort.111" contains the cosine of the matrix A_4

```
        (1.3008916014380032,-0.47788983344969616)
      (-0.19258166557586898,-3.42888850414047042E-003)
     (2.16583885104852554E-002,-0.13446427705269426)
         (1.2644326014253138,-0.59412461049604570)
```

3.5 Power of a matrix

The power of a square matrix A can be found using the spectral theorem as follows. Let us have a matrix A which is diagonalizable with matrix D containing the eigenvalues of A along main diagonal, that is $D = P^{-1}AP$, this implies $A = PDP^{-1}$. Multiplying these equations n times and using the fact that $PP^{-1} = P^{-1}P = \mathbb{I}$, we can see that,

$$A^k = PD^kP^{-1}. \tag{3.20}$$

Here k is any **double precision** real number. Note that, in the above equations if A is a Hermitian matrix then $A^k = UD^kU^\dagger$, where U is a unitary matrix. Power of a matrix finds important applications in very many areas of physics and mathematics.

3.5.1 For a real symmetric matrix

subroutine POWNSR(A,n,k,B)

Parameters

In/Out	Argument	Description
[in/out]	A	A is a real*8 array of dimension (0:n−1,0:n−1), on exit, the columns of A will be orthonormal eigenvectors of A
[in]	n	n (integer) is the dimension of the matrix A and B
[in]	k	k (real*8) is the power to which matrix A is raised
[out]	B	B is a complex*16 array of dimension (0:n−1,0:n−1) which will be A^k

Implementation

```
subroutine POWNSR(A,n,k,B)
implicit none
real*8::A(0:n-1,0:n-1)
complex*16::D(0:n-1,0:n-1),im,B(0:n-1,0:n-1)
integer::n,i
character::JOBZ, UPLO
integer::INFO
real*8::W(0:N-1),WORK(0:3*N-2),k
im=dcmplx(0.d0,1.0d0)
JOBZ='V'
UPLO='U'
call DSYEV(JOBZ, UPLO, N, A, N, W, WORK, 3*N-1, INFO)
D=0.0d0
do i=0,n-1
   if (abs(W(i))<0.0000000000001) then
      W(i)=0.0d0
```

```
      end if
      if (W(i) .lt. 0.0d0) then
          D(i,i)=(abs(W(i))**(k))*(im**(2.0d0*k))
      else
          D(i,i)=W(i)**k
      end if
end do
B=matmul(matmul(A,D),transpose(A))
end subroutine
```

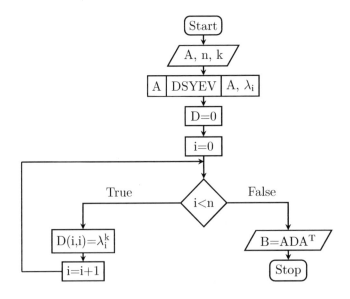

FIGURE 3.11: Flowchart of POWNSR

Example

In this example we are finding the matrix $A_1^{2/3}$.

```
program example
implicit none
real*8::A(0:2,0:2)
complex*16::B(0:2,0:2)
integer::i,j
A(0,:)=(/1,-2,3/)
A(1,:)=(/-2,3,4/)
A(2,:)=(/3,4,5/)
call POWNSR(A,3,2.0d0/3.0d0,B)
do i=0,2,1
    do j=0,2,1
        write(111,*)B(i,j)
    end do
end do
end program
```

The file "fort.111" contains entries of $A_1^{2/3}$.

```
(0.99390455843554326,0.77939749468922248)
(-1.0815701849380792,0.66289918572492201)
(1.4285535685735817,-0.61766176054906730)
(-1.0815701849380794,0.66289918572492201)
(1.8453352141337975,0.56381414288480025)
(1.7513117386048918,-0.52533845811842372)
(1.4285535685735817,-0.61766176054906730)
(1.7513117386048918,-0.52533845811842361)
(2.6240902290194716,0.48948842284474536)
```

3.5.2 For a complex Hermitian matrix

```
subroutine POWNHC(A,n,k,B)
```

Parameters

In/Out	Argument	Description
[in/out]	A	A is a complex*16 array of dimension (0:n−1,0:n−1), on exit, the columns of A will be orthonormal eigenvectors of A
[in]	n	n (integer) is the dimension of the matrix A and B
[in]	k	k (real*8) is the power to which matrix A is raised
[out]	B	B is a complex*16 array of dimension (0:n−1,0:n−1) which will be A^k

Implementation

```
subroutine POWNHC(A,n,k,B)
implicit none
complex*16::A(0:n-1,0:n-1),B(0:n-1,0:n-1),D(0:n-1,0:n-1),im
integer::n,i
character::JOBZ, UPLO
integer::INFO
real*8::W(0:N-1),RWORK(0:3*N-3),k
complex*16::WORK(0:2*N-2)
im=dcmplx(0.d0,1.0d0)
JOBZ='V'
UPLO='U'
call ZHEEV(JOBZ, UPLO, N, A, N, W, WORK, 2*N-1, RWORK,INFO)
D=0.0d0
do i=0,n-1
    if (abs(W(i))<0.0000000000001) then
        W(i)=0.0d0
```

```
    end if
    if (W(i) .lt. 0.0d0) then
        D(i,i)=(abs(W(i))**(k))*(im**(2.0d0*k))
    else
        D(i,i)=W(i)**k
    end if
end do
B=matmul(matmul(A,D),conjg(transpose(A)))
end subroutine
```

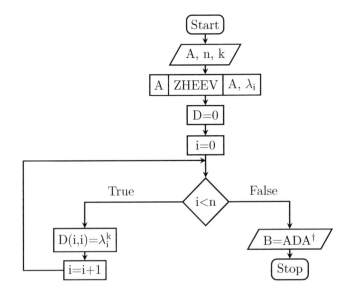

FIGURE 3.12: Flowchart of POWNHC

Example

In this example we are finding the matrix $A_2^{2/3}$.

```
program example
implicit none
complex*16::A(0:1,0:1),B(0:1,0:1)
integer::i,j
A(0,0)=dcmplx(2,0)
A(0,1)=dcmplx(1,-3)
A(1,0)=dcmplx(1,3)
A(1,1)=dcmplx(4,0)
call POWNHC(A,2,2.0d0/3.0d0,B)
do i=0,1,1
    do j=0,1,1
        write(111,*)B(i,j)
    end do
end do
end program
```

The file "fort.111" contains entries of $A_2^{2/3}$.

(1.0422467354458071,0.26180536228272361)
(0.36821119215777487,-1.7111383689854585)
(0.73211406766505538,1.5898374104830317)
(2.1425719952686375,0.14050440378029680)

3.5.3 For a real non-symmetric matrix

subroutine POWNMR(A,n,k,E)

Parameters

In/Out	Argument	Description
[in]	A	A is a real*8 array of dimension (0:n−1,0:n−1)
[in]	n	n (integer) is the dimension of the matrix A and E
[in]	k	k (real*8) is the power to which matrix A is raised
[out]	E	E is a complex*16 array of dimension (0:n−1,0:n−1) which will be A^k

Implementation

```
subroutine POWNMR(A,n,k,E)
implicit none
integer::n,i,j
complex*16::E(0:n-1,0:n-1),D(0:n-1,0:n-1)
complex*16::B(0:n-1,0:n-1),C(0:n-1,0:n-1)
character*3::JOBVL, JOBVR
integer::INFO
complex*16::W(0:N-1),VL(0:N-1,0:N-1),VR(0:N-1,0:N-1)
complex*16::WORK(0:2*N-1)
real*8::RWORK(0:2*N-1),k,A(0:n-1,0:n-1)
JOBVL='N'
JOBVR='V'
do i=0,n-1
    do j=0,n-1
        C(i,j)=dcmplx(A(i,j),0.0d0)
    end do
end do
call ZGEEV( JOBVL, JOBVR, N, C, N, W, VL, N, &
    VR, N, WORK, 2*N, RWORK, INFO )
D=0.0d0
do i=0,n-1
    if (abs(W(i))<0.0000000000001) then
        W(i)=0
```

```
      end if
      D(i,i)=W(i)**(k)
end do
call MATINVMC(n,VR,B)
E=matmul(matmul(VR,D),B)
end subroutine
```

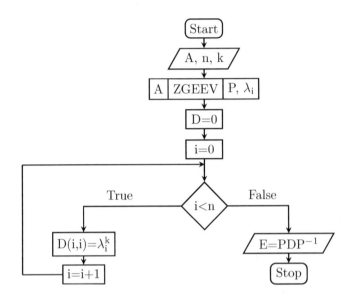

FIGURE 3.13: Flowchart of POWNMR

Example

In this example we are finding the matrix A_3^2.

```
program example
implicit none
real*8::A(0:2,0:2)
complex*16::B(0:2,0:2)
integer::i,j
A(0,:)=(/1.0,-2.0,3.0/)
A(1,:)=(/2.0,3.0,5.0/)
A(2,:)=(/-4.0,4.0,5.0/)
call POWNMR(A,3,2.0d0,B)
do i=0,2,1
    write(111,*)real(B(i,:))
end do
end program
```

The file "fort.111" contains entries of A_3^2.

-14.999999999999995	4.0000000000000062	8.0000000000000124
-12.000000000000009	25.000000000000014	46.000000000000007
-16.000000000000004	39.999999999999979	32.999999999999993

3.5.4 For a complex non-symmetric matrix

---
subroutine POWNMC(A,n,k,E)

Parameters

In/Out	Argument	Description
[in/out]	A	A is a complex*16 array of dimension (0:n−1,0:n−1), on exit, A will be overwritten
[in]	n	n (integer) is the dimension of the matrix A and E
[in]	k	k (real*8) is the power to which matrix A is raised
[out]	E	E is a complex*16 array of dimension (0:n−1,0:n−1) which will be A^k

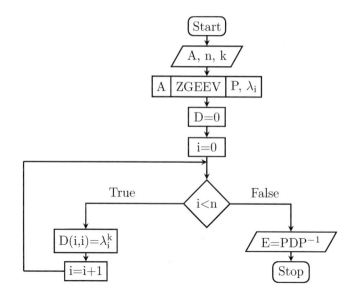

FIGURE 3.14: Flowchart of POWNMC

Implementation

```fortran
subroutine POWNMC(A,n,k,E)
implicit none
integer::n,i
complex*16::A(0:n-1,0:n-1),B(0:n-1,0:n-1)
complex*16::D(0:n-1,0:n-1),E(0:n-1,0:n-1)
```

```
character*3::JOBVL, JOBVR
integer::INFO
complex*16::W(0:N-1),VL(0:N-1,0:N-1),VR(0:N-1,0:N-1)
complex*16::WORK(0:2*N-1)
real*8::RWORK(0:2*N-1),k
JOBVL='N'
JOBVR='V'
call ZGEEV(JOBVL, JOBVR, N, A, N, W, VL, N, VR, &
        N, WORK, 2*N, RWORK, INFO)
D=0.0d0
do i=0,n-1
if (abs(W(i))<0.0000000000001) then
W(i)=0
end if
D(i,i)=W(i)**(k)
end do
call MATINVMC(n,VR,B)
 E=matmul(matmul(VR,D),B)
end subroutine
```

Example

In this example we are finding the matrix A_4^2.

```
program example
implicit none
complex*16::A(0:1,0:1),B(0:1,0:1)
integer::i
A(0,0)=dcmplx(2,0)
A(0,1)=dcmplx(1,-3)
A(1,0)=dcmplx(2,1)
A(1,1)=dcmplx(4,0)
call POWNMC(A,2,2.0d0,B)
do i=0,1,1
    write(111,*)B(i,:)
end do
end program
```

The file "fort.111" contains entries of A_4^2.

```
(9.0000000000000000,-4.9999999999999964)
(6.0000000000000036,-17.999999999999993)
(12.000000000000000,6.0000000000000018)
(21.000000000000000,-5.0000000000000000)
```

3.6   Trace of a power of the matrix

Trace of a diagonalizable matrix can be written in terms of the eigenvalues of a matrix as follows. Let us have a $n \times n$ matrix A, which has eigenvalues λ_i. Trace of the matrix, can be written as,

$$Tr(A) = \lambda_1 + \lambda_2 + \lambda_3 + \cdots + \lambda_n = \sum_{i=1}^{n} \lambda_i. \qquad (3.21)$$

Similarly using Eq. [3.20], we can get the trace of the kth power of the matrix A as,

$$Tr(A^k) = \lambda_1^k + \lambda_2^k + \lambda_3^k + \cdots + \lambda_n^k = \sum_{i=1}^{n} \lambda_i^k. \qquad (3.22)$$

3.6.1   For a real symmetric matrix

subroutine TRPOWNSR(A,n,k,trace)

Parameters

In/Out	Argument	Description
[in]	n	n (integer) is the dimension of the matrix A
[in/out]	A	A is a real*8 array of the dimension (0:n−1,0:n−1), on exit A will be overwritten
[in]	k	k (real*8) is the power to which the matrix A is raised
[out]	trace	trace (complex*16) is the trace of the matrix A^k

Implementation

```
subroutine TRPOWNSR(A,n,k,trace)
implicit none
real*8::A(0:n-1,0:n-1)
complex*16::trace,im
integer::n,i
character::JOBZ, UPLO
integer::INFO
real*8::W(0:N-1),WORK(0:3*N-2),k
JOBZ='N'
UPLO='U'
im=dcmplx(0.0d0,1.0d0)
call DSYEV(JOBZ, UPLO, N, A, N, W, WORK, 3*N-1, INFO)
trace=0.0d0
do i=0,n-1
    if (abs(W(i))<0.0000000000001) then
        W(i)=0.0d0
```

```
      end if
      if (W(i) .lt. 0.0d0) then
          trace=trace+(abs(W(i))**(k))*(im**(2.0d0*k))
      else
          trace=trace+W(i)**k
      end if

end do
end subroutine
```

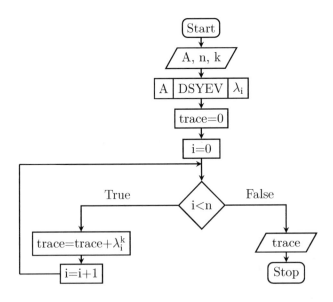

FIGURE 3.15: Flowchart of TRPOWNSR

Example

In this example we are finding the trace of the matrix $A_1^{2/3}$.

```
program example
implicit none
real*8::A(0:2,0:2)
complex*16::trace
integer::i
A(0,:)=(/1,-2,3/)
A(1,:)=(/-2,3,4/)
A(2,:)=(/3,4,5/)
call TRPOWNSR(A,3,2.0d0/3.0d0,trace)
write(111,*)trace
end program
```

The file "fort.111" contains the trace of $A_1^{2/3}$.

(5.4633300015888153,1.8327000604187680)

3.6.2 For a complex Hermitian matrix

subroutine TRPOWNHC(A,n,k,trace)

Parameters

In/Out	Argument	Description
[in]	n	n (integer) is the dimension of the matrix A
[in/out]	A	A is a complex*16 array of dimension $(0{:}n{-}1,0{:}n{-}1)$, on exit A will be overwritten
[in]	k	k (real*8) is the power to which the matrix A is raised
[out]	trace	trace (complex*16) is the trace of the matrix A^k

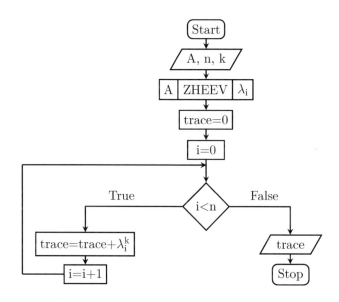

FIGURE 3.16: Flowchart of TRPOWNHC

Implementation

```
subroutine TRPOWNHC(A,n,k,trace)
implicit none
complex*16::A(0:n-1,0:n-1),trace,im
integer::n,i
character::JOBZ, UPLO
integer::INFO
real*8::W(0:N-1),RWORK(0:3*N-3),k
complex*16::WORK(0:2*N-2)
JOBZ='N'
UPLO='U'
```

```
im=dcmplx(0.0d0,1.0d0)
call ZHEEV(JOBZ, UPLO, N, A, N, W, WORK, 2*N-1, RWORK, INFO)
trace=0.0d0
do i=0,n-1
    if (abs(W(i))<0.0000000000001) then
        W(i)=0.0d0
    end if
    if (W(i) .lt. 0.0d0) then
        trace=trace+(abs(W(i))**(k))*(im**(2.0d0*k))
    else
        trace=trace+W(i)**k
    end if
end do
end subroutine
```

Example

In this example we are finding the trace of the matrix $A_2^{2/3}$.

```
program example
implicit none
complex*16::A(0:1,0:1),trace
integer::i,j
A(0,0)=dcmplx(2,0)
A(0,1)=dcmplx(1,-3)
A(1,0)=dcmplx(1,3)
A(1,1)=dcmplx(4,0)
call TRPOWNHC(A,2,2.0d0/3.0d0,trace)
write(111,*)trace
end program
```

The file "fort.111" contains the trace of $A_2^{2/3}$.

$$(3.1848187307144444,0.40230976606302044)$$

3.6.3   For a real non-symmetric matrix

```
subroutine TRPOWNMR(A,n,k,trace)
```

Parameters

In/Out	Argument	Description
[in]	n	n (integer) is the dimension of the matrix A
[in]	A	A is real*8 array of dimension (0:n−1,0:n−1), on exit A will be overwritten.
[in]	k	k (real*8) is the power to which the matrix A is raised
[out]	trace	trace (complex*16) is the trace of the matrix A^k

Implementation

```
subroutine TRPOWNMR(A,n,k,trace)
implicit none
integer::n,i
real*8::k
complex*16::trace,o
character::JOBVL, JOBVR
integer::INFO
real*8::WR(0:N-1),WI(0:N-1),VL(0:N-1,0:N-1)
real*8::VR(0:N-1,0:N-1),WORK(0:4*N-1),A(0:N-1,0:N-1)
JOBVL='N'
JOBVR='N'
call DGEEV( JOBVL, JOBVR, N, A, N, WR, WI, &
              VL, N, VR,N, WORK, 4*N, INFO )
trace=0.0d0
do i=0,n-1,1
    o=dcmplx(WR(i),WI(i))
    if (abs(o)<0.0000000000001) then
        o=0.0d0
    end if
trace=trace+o**k
end do
end subroutine
```

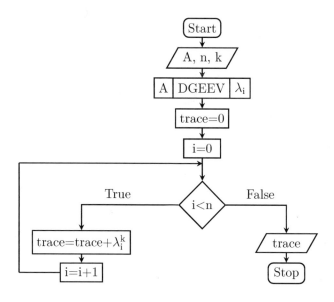

FIGURE 3.17: Flowchart of TRPOWNMR

Example

In this example we are finding the trace of the matrix A_3^2.

```
program example
```

```
implicit none
real*8::A(0:2,0:2)
complex*16::trace
integer::i,j
A(0,:)=(/1.0,-2.0,3.0/)
A(1,:)=(/2.0,3.0,5.0/)
A(2,:)=(/-4.0,4.0,5.0/)
call TRPOWNMR(A,3,2.0d0,trace)
write(111,*)trace
end program
```

The file "fort.111" contains the trace of A_3^2.

(43.000000000000014,0.0000000000000000)

3.6.4   For a complex non-symmetric matrix

```
subroutine TRPOWNMC(A,n,k,trace)
```

Parameters

In/Out	Argument	Description
[in]	n	n (integer) is the dimension of the matrix A
[in]	A	A is a complex*16 array of dimension (0:n−1,0:n−1), on exit A will be overwritten
[in]	k	k (real*8) is the power to which the matrix A is raised
[out]	trace	trace (complex*16) is the trace of the matrix A^k

Implementation

```
subroutine TRPOWNMC(A,n,k,trace)
implicit none
integer::n,i
real*8::k
complex*16::trace,o
character::JOBVL, JOBVR
integer::INFO
complex*16::W(0:N-1),VL(0:N-1,0:N-1)
complex*16::VR(0:N-1,0:N-1),WORK(0:2*N-1),A(0:N-1,0:N-1)
real*8::RWORK(0:2*N-1)
JOBVL='N'
JOBVR='N'
call ZGEEV(JOBVL, JOBVR, N, A, N, W, VL, N, &
              VR, N, WORK, 2*N, RWORK, INFO)
trace=0.0d0
```

```
do i=0,n-1,1
    if (abs(W(i))<0.0000000000001) then
        W(i)=0.0d0
    end if
o=W(i)
trace=trace+o**k
end do
end subroutine
```

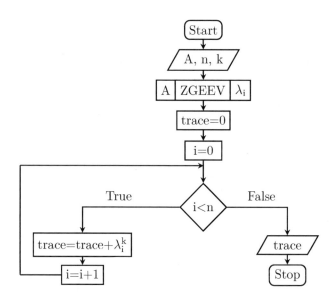

FIGURE 3.18: Flowchart of POWNMC

Example

In this example we are finding the trace matrix A_4^2.

```
program example
implicit none
complex*16::A(0:1,0:1),trace
A(0,0)=dcmplx(2,0)
A(0,1)=dcmplx(1,-3)
A(1,0)=dcmplx(2,1)
A(1,1)=dcmplx(4,0)
call TRPOWNMC(A,2,2.0d0,trace)
write(111,*)trace
end program
```

The file "fort.111" contains the trace of matrix A_4^2.

```
(30.000000000000000,-9.9999999999999964)
```

3.7   Determinant of a Matrix

Let A be a $n \times n$ matrix and let λ_i be its eigenvalues. Then the determinant of matrix A can be given as the product of its eigenvalues as below,

$$\det A = \lambda_1 \times \lambda_2 \times \cdots \times \lambda_n = \prod_{i=1}^{n} \lambda_i. \tag{3.23}$$

Note that, even if one of the eigenvalues of matrix A will be zero, then the determinant will be zero.

3.7.1   For a real symmetric matrix

subroutine DETSR(A,n,det)

Parameters

In/Out	Argument	Description
[in]	n	n (integer) is the dimension of the matrix A
[in]	A	A is a real*8 array of dimension (0:n−1,0:n−1), on exit A will be overwritten
[out]	det	det (real*8) is the determinant of the matrix A

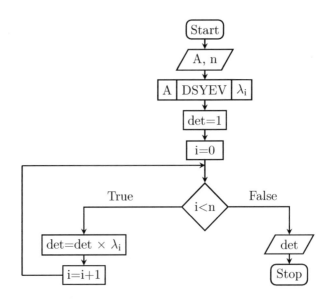

FIGURE 3.19: Flowchart of DETSR

Implementation

```
subroutine DETSR(A,n,det)
implicit none
real*8::A(0:n-1,0:n-1),det
integer::n,i
character::JOBZ, UPLO
integer::INFO
real*8::W(0:N-1),WORK(0:3*N-2)
JOBZ='N'
UPLO='U'
call DSYEV(JOBZ, UPLO, N, A, N, W, WORK, 3*N-1, INFO)
det=1.0d0
do i=0,n-1
    det=det*W(i)
end do
end subroutine
```

Example

In this example we are finding the determinant of the matrix A_1.

```
program example
implicit none
real*8::A(0:2,0:2),det
integer::i
A(0,:)=(/1,-2,3/)
A(1,:)=(/-2,3,4/)
A(2,:)=(/3,4,5/)
call DETSR(A,3,det)
write(111,*)det
end program
```

The file "fort.111" contains the determinant of the matrix A_1.

```
-96.000000000000000
```

3.7.2 For a complex Hermitian matrix

```
subroutine DETHC(A,n,det)
```

Parameters

In/Out	Argument	Description
[in]	n	n (integer) is the dimension of the matrix A
[in]	A	A is a complex*16 array of dimension (0:n−1,0:n−1), on exit A will be overwritten

In/Out	Argument	Description
[out]	det	det (real*8) is the determinant of the matrix A

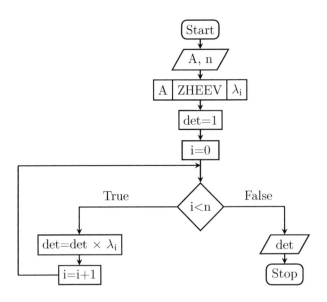

FIGURE 3.20: Flowchart of DETHC

Implementation

```
subroutine DETHC(A,n,det)
implicit none
complex*16::A(0:n-1,0:n-1)
integer::n,i,INFO
character::JOBZ, UPLO
real*8::W(0:N-1),RWORK(0:3*N-3),det
complex*16::WORK(0:2*N-2)
JOBZ='N'
UPLO='U'
call ZHEEV(JOBZ, UPLO, N, A, N, W, WORK, 2*N-1, RWORK,INFO)
det=1.0d0
do i=0,n-1
   det=det*W(i)
end do
end subroutine
```

Example

In this example we are finding the determinant of the matrix A_2.

```
program example
implicit none
```

```
complex*16::A(0:1,0:1)
real*8::det
integer::i,j
A(0,0)=dcmplx(2,0)
A(0,1)=dcmplx(1,-3)
A(1,0)=dcmplx(1,3)
A(1,1)=dcmplx(4,0)
call DETHC(A,2,det)
write(111,*)det
end program
```

The file "fort.111" contains the determinant of the matrix A_2.

```
-2.0000000000000013
```

3.7.3   For a real non-symmetric matrix

```
subroutine DETMR(A,n,det)
```

Parameters

In/Out	Argument	Description
[in]	n	n (integer) is the dimension of the matrix A
[in]	A	A is a real*8 array of dimension $(0:n-1,0:n-1)$, on exit A will be overwritten
[out]	det	det (real*8) is the determinant of the matrix A

Implementation

```
subroutine DETMR(A,n,det)
implicit none
integer::n,i
real*8::det
complex*16::det1,o
character::JOBVL, JOBVR
integer::INFO
real*8::WR(0:N-1),WI(0:N-1),VL(0:N-1,0:N-1)
real*8::VR(0:N-1,0:N-1),WORK(0:4*N-1),A(0:N-1,0:N-1)
JOBVL='N'
JOBVR='N'
call DGEEV(JOBVL, JOBVR, N, A, N, WR, WI, VL, &
              N, VR,N, WORK, 4*N, INFO)
det1=dcmplx(1.0d0,0.d0)
do i=0,n-1,1
   o=dcmplx(WR(i),WI(i))
   det1=det1*o
```

```
end do
det=real(det1)
end subroutine
```

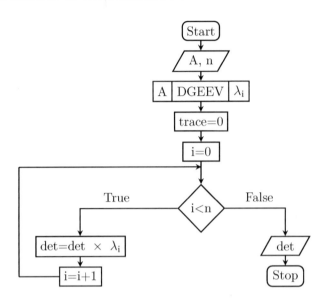

FIGURE 3.21: Flowchart of DETMR

Example

In this example we are finding the determinant of the matrix A_3.

```
program example
implicit none
real*8::A(0:2,0:2),det
integer::i,j
A(0,:)=(/1.0,-2.0,3.0/)
A(1,:)=(/2.0,3.0,5.0/)
A(2,:)=(/-4.0,4.0,5.0/)
call DETMR(A,3,det)
write(111,*)det
end program
```

The file "fort.111" contains the determinant of the matrix A_3.

```
 114.99999999999989
```

3.7.4   For a complex non-symmetric matrix

```
subroutine DETMC(A,n,det)
```

Parameters

In/Out	Argument	Description
[in]	n	n (integer) is the dimension of the matrix A
[in]	A	A is a complex*16 array of dimension $(0:n-1,0:n-1)$, on exit A will be overwritten
[out]	det	det (complex*16) is the determinant of the matrix A

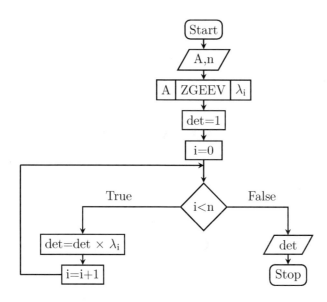

FIGURE 3.22: Flowchart of DETMC

Implementation

```
subroutine DETMC(A,n,det)
implicit none
integer::n,i
complex*16::det
character::JOBVL, JOBVR
integer::INFO
complex*16::W(0:N-1),VL(0:N-1,0:N-1)
complex*16::VR(0:N-1,0:N-1),WORK(0:2*N-1),A(0:N-1,0:N-1)
real*8::RWORK(0:2*N-1)
JOBVL='N'
JOBVR='N'
call ZGEEV(JOBVL, JOBVR, N, A, N, W, VL, &
                    N, VR, N, WORK, 2*N, RWORK, INFO)
det=dcmplx(1.0d0,0.d0)
do i=0,n-1,1
   det=det*W(i)
```

```
end do
end subroutine
```

Example

In this example we are finding the determinant of the matrix A_4.

```
program example
implicit none
complex*16::A(0:1,0:1),det
A(0,0)=dcmplx(2,0)
A(0,1)=dcmplx(1,-3)
A(1,0)=dcmplx(2,1)
A(1,1)=dcmplx(4,0)
call DETMC(A,2,det)
write(111,*)det
end program
```

The file "fort.111" contains the determinant of the matrix A_4.

```
(2.9999999999999960,4.9999999999999973)
```

3.8   Trace norm of a matrix

Let A be a $n \times n$ matrix, note that even if matrix A is not diagonalizable, $A^\dagger A$ is a positive Hermitian matrix and hence diagonalizable by a unitary transformation. We then define the trace norm of matrix [12, 22] A as

$$\| A \| = Tr\left(\sqrt{A^\dagger A}\right). \tag{3.24}$$

In the above equation, the square root is uniquely determined by spectral theorem as given by Eq. [3.19]. Trace norm has applications in the calculation of logarithmic negativity of a state, computation of fidelity between two density matrices and so on.

3.8.1   For a real matrix

```
subroutine TRNORMMR(A,n,trace_norm)
```

Parameters

In/Out	Argument	Description
[in]	n	n (integer) is the dimension of the matrix A
[in]	A	A is a real*8 array of dimension $(0:n-1,0:n-1)$
[out]	trace_norm	trace_norm (real*8) is the trace norm of the matrix A

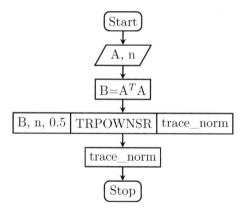

FIGURE 3.23: Flowchart of TRNORMMR

Implementation

```
subroutine TRNORMMR(A,n,trace_norm)
implicit none
integer::n
real*8::A(0:N-1,0:N-1),B(0:N-1,0:N-1),trace_norm
complex*16::trace
B=matmul(transpose(A),A)
call TRPOWNSR(B,n,0.5d0,trace)
trace_norm=real(trace)
end subroutine
```

Example

In this example we are finding the trace norm of the matrix A_3.

```
program example
implicit none
real*8::A(0:2,0:2),B(0:2,0:2),trace_norm
A(0,:)=(/1.0,-2.0,3.0/)
A(1,:)=(/2.0,3.0,5.0/)
A(2,:)=(/-4.0,4.0,5.0/)
call TRNORMMR(A,3,trace_norm)
write(111,*)trace_norm
end program
```

The file "fort.111" contains the trace norm of the matrix A_3.

```
16.358044723373784
```

3.8.2 For a complex matrix

```
subroutine TRNORMMC(A,n,trace_norm)
```

Parameters

In/Out	Argument	Description
[in]	n	n (integer) is the dimension of the matrix A
[in]	A	A is a complex*16 array of dimension $(0{:}n{-}1,0{:}n{-}1)$
[out]	trace_norm	trace_norm (real*8) is the trace norm of the matrix A

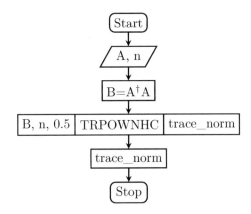

FIGURE 3.24: Flowchart of TRNORMMC

Implementation

```
subroutine TRNORMMC(A,n,trace_norm)
implicit none
integer::n
real*8::trace_norm
complex*16::A(0:N-1,0:N-1),B(0:N-1,0:N-1),trace
B=matmul(conjg(transpose(A)),A)
call TRPOWNHC(B,n,0.5d0,trace)
trace_norm=real(trace)
end subroutine
```

Example

In this example we are finding the trace norm of the matrix A_4.

```
program example
implicit none
complex*16::A(0:1,0:1),B(0:1,0:1)
real*8::trace_norm
A(0,0)=dcmplx(2,0)
A(0,1)=dcmplx(1,-3)
A(1,0)=dcmplx(2,1)
A(1,1)=dcmplx(4,0)
```

```
call TRNORMMC(A,2,trace_norm)
write(111,*)trace_norm
end program
```

The file "fort.111" contains the trace norm of the matrix A_4.

6.8309518948453007

3.9   Hilbert-Schmidt norm of a matrix

Let A be an $n \times n$ matrix, then the Hilbert-Schmidt norm [12,22] of matrix A can be defined as,

$$\| A \|_{HS} = \sqrt{Tr(A^\dagger A)}. \tag{3.25}$$

3.9.1   For a real matrix

```
subroutine HISCNORMMR(A,n,HS_norm)
```

Parameters

In/Out	Argument	Description
[in]	n	n (integer) is the dimension of the matrix A
[in]	A	A is a real*8 array of dimension (0:n−1,0:n−1)
[out]	HS_norm	HS_norm (real*8) is the Hilbert-Schmidt norm of the matrix A

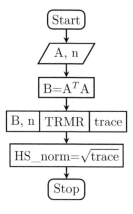

FIGURE 3.25: Flowchart of HISCNORMMR

Implementation

```
subroutine HISCNORMMR(A,n,HS_norm)
implicit none
integer::n
real*8::trace,HS_norm,A(0:n-1,0:n-1),B(0:n-1,0:n-1)
B=matmul(transpose(A),A)
call TRMR(B,n,trace)
HS_norm=sqrt(trace)
end subroutine
```

Example

In this example we are finding the Hilbert-Schmidt norm of the matrix A_3.

```
program example
implicit none
real*8::A(0:2,0:2),B(0:2,0:2),HS_norm
A(0,:)=(/1.0,-2.0,3.0/)
A(1,:)=(/2.0,3.0,5.0/)
A(2,:)=(/-4.0,4.0,5.0/)
call HISCNORMMR(A,3,HS_norm)
write(111,*)HS_norm
end program
```

The file "fort.111" contains the Hilbert-Schmidt norm of the matrix A_3.

```
10.440306508910547
```

3.9.2 For a complex matrix

```
subroutine HISCNORMMC(A,n,HS_norm)
```

Parameters

In/Out	Argument	Description
[in]	n	n (integer) is the dimension of the matrix A
[in]	A	A is a complex*16 array of dimension $(0:n-1,0:n-1)$
[out]	HS_norm	HS_norm (real*8) is the Hilbert-Schmidt norm of the matrix A

Implementation

```
subroutine HISCNORMMC(A,n,HS_norm)
implicit none
integer::n
```

```
complex*16::A(0:n-1,0:n-1),B(0:n-1,0:n-1),trace
real*8::HS_norm
B=matmul(conjg(transpose(A)),A)
call TRMC(B,n,trace)
HS_norm=sqrt(abs(trace))
end subroutine
```

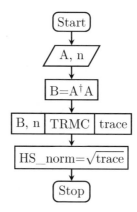

FIGURE 3.26: Flowchart of HISCNORMMC

Example

In this example we are finding the Hilbert-Schmidt norm of the matrix A_4.

```
program example
implicit none
complex*16::A(0:1,0:1),B(0:1,0:1)
real*8::HS_norm
A(0,0)=dcmplx(2,0)
A(0,1)=dcmplx(1,-3)
A(1,0)=dcmplx(2,1)
A(1,1)=dcmplx(4,0)
call HISCNORMMC(A,2,HS_norm)
write(111,*)HS_norm
end program
```

The file "fort.111" contains the Hilbert-Schmidt norm of the matrix A_4.

```
5.9160797830996161
```

3.10 Absolute value of a matrix

Let A be a $n \times n$ matrix, then the absolute value of matrix [23] A can be defined as

$$absA = \sqrt{A^\dagger A}, \tag{3.26}$$

here the square root is the positive square root.

3.10.1   For a real matrix

---

subroutine ABSMR(A,n,B)

Parameters

In/Out	Argument	Description
[in]	n	n (integer) is the dimension of the matrices A and B
[in]	A	A is a real*8 array of dimension $(0:n-1,0:n-1)$
[out]	B	B is a real*8 array of dimension $(0:n-1,0:n-1)$ which is the absolute value of matrix A

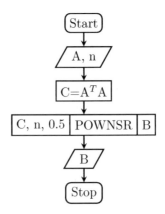

FIGURE 3.27: Flowchart of ABSMR

Implementation

```
subroutine ABSMR(A,n,B)
implicit none
real*8::A(0:n-1,0:n-1),C(0:n-1,0:n-1),B(0:n-1,0:n-1)
complex*16::D(0:n-1,0:n-1)
integer::n
C=matmul(transpose(A),A)
call POWNSR(C,n,0.5d0,D)
B=real(D)
end subroutine
```

Example

In this example we are finding the absolute value of the matrix A_3.

```
program example
implicit none
```

```
real*8::A(0:2,0:2),B(0:2,0:2)
integer::i
A(0,:)=(/1.0,-2.0,3.0/)
A(1,:)=(/2.0,3.0,5.0/)
A(2,:)=(/-4.0,4.0,5.0/)
call ABSMR(A,3,B)
do i=0,2,1
    write(111,*)B(i,:)
end do
end program
```

The file "fort.111" contains entries of the absolute value of the matrix A_3.

4.4001628564621269	-1.2321665557606594	-0.34688963010726387
-1.2321665557606591	4.6660871038599927	2.3894343929173654
-0.34688963010726387	2.3894343929173649	7.2917947630516613

3.10.2 For a complex matrix

```
subroutine ABSMC(A,n,B)
```

Parameters

In/Out	Argument	Description
[in]	n	n (integer) is the dimension of the matrices A and B
[in]	A	A is a complex*16 array of dimension (0:n−1,0:n−1)
[out]	B	B is a complex*16 array of dimension (0:n−1,0:n−1) which is the absolute value of matrix A

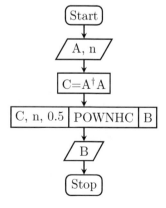

FIGURE 3.28: Flowchart of ABSMC

Implementation

```
subroutine ABSMC(A,n,B)
implicit none
complex*16::A(0:n-1,0:n-1),B(0:n-1,0:n-1),C(0:N-1,0:N-1)
integer::n
C=matmul(conjg(transpose(A)),A)
call POWNHC(C,n,0.5d0,B)
end subroutine
```

Example

In this example we are finding the absolute value of the matrix A_4.

```
program example
implicit none
complex*16::A(0:1,0:1),B(0:1,0:1)
integer::i,j
A(0,0)=dcmplx(2,0)
A(0,1)=dcmplx(1,-3)
A(1,0)=dcmplx(2,1)
A(1,1)=dcmplx(4,0)
call ABSMC(A,2,B)
do i=0,1,1
    do j=0,1,1
        write(111,*)B(i,j)
    end do
end do
end program
```

The file "fort.111" contains the entries of the absolute value of the matrix A_4.

```
(2.1711398532958297,-2.69566084788785821E-018)
    (1.4639248166197880,-1.4639248166197878)
    (1.4639248166197882,1.4639248166197882)
    (4.6598120415494702,0.0000000000000000)
```

3.11   The Hilbert-Schmidt inner product between two matrices

Let A and B be two $n \times n$ matrices or operators, then the Hilbert-Schmidt inner product [24] between them can be defined as,

$$\langle A, B \rangle_{HS} = Tr(A^\dagger B) \tag{3.27}$$

3.11.1   For real matrices

```
subroutine HISCIPMR(n,A,B,HS_inp)
```

Parameters

In/Out	Argument	Description
[in]	n	n (integer) is the dimension of matrices A and B
[in]	A	A is a real*8 array of dimension (0:n−1,0:n−1)
[in]	B	B is a real*8 array of dimension (0:n−1,0:n−1)
[out]	HS_inp	HS_inp (real*8) is the Hilbert-Schmidt inner product between matrices A and B

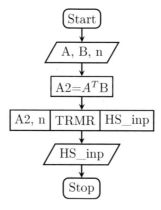

FIGURE 3.29: Flowchart of HISCIPMR

Implementation

```
subroutine HISCIPMR(n,A,B,HS_inp)
integer::n
real*8::A(0:n-1,0:n-1),HS_inp,A2(0:n-1,0:n-1),B(0:n-1,0:n-1)
A2=matmul(transpose(A),B)
call TRMR(A2,n,HS_inp)
end subroutine
```

Example

In this example we are finding the Hilbert-Schmidt inner product between matrices A_1 and A_3.

```
program example
implicit none
real*8::A(0:2,0:2),B(0:2,0:2),HS_inp
integer::i
A(0,:)=(/1,-2,3/)
A(1,:)=(/-2,3,4/)
A(2,:)=(/3,4,5/)
```

```
B(0,:)=(/1.0,-2.0,3.0/)
B(1,:)=(/2.0,3.0,5.0/)
B(2,:)=(/-4.0,4.0,5.0/)
call HISCIPMR(3,A,B,HS_inp)
write(111,*)HS_inp
end program
```

The file "fort.111" contains the Hilbert-Schmidt inner product between matrices A_1 and A_3.

```
   68.000000000000000
```

3.11.2　For complex matrices

```
subroutine HISCIPMC(n,A,B,HS_inp)
```

Parameters

In/Out	Argument	Description
[in]	n	n (integer) is the dimension of matrices A and B
[in]	A	A is a complex*16 array of dimension $(0:n-1,0:n-1)$
[in]	B	B is a complex*16 array of dimension $(0:n-1,0:n-1)$
[out]	HS_inp	HS_inp (complex*16) is the Hilbert schmidt inner product between matrices A and B

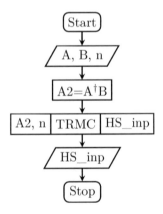

FIGURE 3.30: Flowchart of HISCIPMC

Implementation

```
subroutine HISCIPMC(n,A,B,HS_inp)
```

```
integer::n
complex*16::A(0:n-1,0:n-1),HS_inp,A2(0:n-1,0:n-1),B(0:n-1,0:n-1)
A2=matmul(transpose(conjg(A)),B)
call TRMC(A2,n,HS_inp)
end subroutine
```

Example

In this example we are finding the Hilbert-Schmidt inner product between matrices A_2 and A_4.

```
program example
implicit none
complex*16::A(0:1,0:1),B(0:1,0:1),HS_inp
A(0,0)=dcmplx(2,0)
A(0,1)=dcmplx(1,-3)
A(1,0)=dcmplx(1,3)
A(1,1)=dcmplx(4,0)
B(0,0)=dcmplx(2,0)
B(0,1)=dcmplx(1,-3)
B(1,0)=dcmplx(2,1)
B(1,1)=dcmplx(4,0)
call HISCIPMC(2,A,B,HS_inp)
write(111,*)HS_inp
end program
```

The file "fort.111" contains the Hilbert-Schmidt inner product between matrices A_2 and A_4.

$$(35.000000000000000,-5.0000000000000000)$$

3.12 Gram-Schmidt orthogonalization

Let there be a linearly independent non-orthogonal basis set $|w\rangle : \{|w_1\rangle, |w_2\rangle, \cdots, |w_m\rangle\}$ (i.e. $\langle w_i|w_j\rangle \neq \delta_{ij}$) in some vector space V where an inner product is defined. There is a useful method known as 'Gram-Schmidt' procedure [11,25,26] which can be used to generate a set of orthonormal vectors $|v\rangle : \{|v_1\rangle, |v_2\rangle, \cdots, |v_m\rangle\}$ from the non-orthonormal set $|w\rangle$. Define initially,

$$|v_1\rangle = \frac{|w_1\rangle}{\||w_1\rangle\|}. \tag{3.28}$$

For $1 \leq k \leq m-1$ define $|v_{k+1}\rangle$ inductively by,

$$|v_{k+1}\rangle \equiv \frac{|w_{k+1}\rangle - \sum_{i=1}^{k}\langle v_i|w_{k+1}\rangle|v_i\rangle}{\||w_{k+1}\rangle - \sum_{i=1}^{k}\langle v_i|w_{k+1}\rangle|v_i\rangle\|}. \tag{3.29}$$

We can illustrate this process by taking an example of vectors on a two-dimensional real vector space, which can be represented on a plane, consider vectors w_1 and w_2 inclined at an angle $\theta \neq \pi/2$. Initially, consider $|v_1\rangle = |w_1\rangle$, Now take the second vector $|w_2\rangle$ from the set of non-orthogonal basis and project it along the direction of $|v_1\rangle$ using the projection operator $P_1 = |v_1\rangle\langle v_1|$ as $P_1|w_2\rangle = |v_1\rangle\langle v_1|w_2\rangle$. Now subtract $P_1|w_2\rangle$ from $|w_2\rangle$ respecting the triangle law of addition of vectors to get $|v_2\rangle = |w_2\rangle - P_1|w_2\rangle = |w_2\rangle - |v_1\rangle\langle v_1|w_2\rangle$, at this stage $|v_1\rangle$ and $|v_2\rangle$ are orthogonal, however to normalize them, we just divide it by the norm of $|v_2\rangle = |\,\||w_2\rangle - |v_1\rangle\langle v_1|w_2\rangle\|$ to obtain the orthonormal pair $|v_1\rangle$ and $|v_2\rangle = \dfrac{|w_2\rangle - |v_1\rangle\langle v_1|w_2\rangle}{\|\,|w_2\rangle - |v_1\rangle\langle v_1|w_2\rangle\|}$ such that $\langle v_1|v_2\rangle = \delta_{12}$ now it is easy to see that, when we extend this to more than 2 vectors, we get the formula as given in Eq. [3.29]. The Gram-Schmidt procedure has many applications like generating a set of orthonormal vectors from a non-orthonormal set, checking whether a set of vectors are linearly dependent or not, in fact the matrix Q which we have in the QR decomposition of a matrix can also be constructed using the Gram-Schmidt procedure.

3.12.1   For real vectors

subroutine GSOMR(n,k,wi,vi)

Parameters

In/Out	Argument	Description
[in]	n	n (integer) is the number of rows in the matrix wi
[in]	k	k (integer) is the number of columns in the matrix wi
[in]	wi	wi is a real*8 array of dimension (0:n−1,0:k−1), whose columns are the linearly independent non-orthonormal set of vectors
[out]	vi	vi is a real*8 array of dimension (0:n−1,0:k−1) which stores the k vectors as columns of dimension n which is the orthonormalized set of vectors

Implementation

```
subroutine GSOMR(n,k,wi,vi)
implicit none
integer::n,k,i,j
real*8::wi(0:n-1,0:k-1),vi(0:n-1,0:k-1),c
real*8,dimension(0:n-1)::u1,u2,u3,u4,u5
vi=0.0d0
u1=wi(:,0)
call NORMALIZEVR(u1,n)
vi(:,0)=u1
do i=1,k-1
    u2=wi(:,i)
    u3=0.0d0
    u4=0.0d0
    do j=0,i-1
        u5=vi(:,j)
```

```
        call IPVR(u5,u2,n,c)
        u3=u3+c*u5
    end do
    u4=u2-u3
    call NORMALIZEVR(u4,n)
    vi(:,i)=u4
end do
end subroutine
```

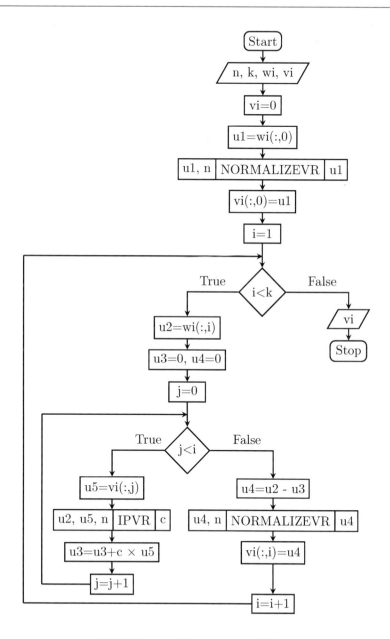

FIGURE 3.31: Flowchart of GSOMR

Example

In this example we are doing the Gram-Schmidt orthogonalization of the matrix A whose columns are the linearly independent non-orthogonal vectors.

$$A = \begin{pmatrix} 1 & 2 & 3 & 4 \\ 2 & 33 & 4 & 5 \\ 3 & 4 & 53 & 6 \\ 4 & 5 & 6 & 73 \end{pmatrix}. \tag{3.30}$$

```
program example
implicit none
integer::i
real*8::m1(0:3,0:3),m2(0:3,0:3)
m1(0,:)=(/1,2,3,4/)
m1(1,:)=(/2,33,4,5/)
m1(2,:)=(/3,4,53,6/)
m1(3,:)=(/4,5,6,73/)
call GSOMR(4,4,m1,m2)
do i=0,3
    write(111,*)real(m2(i,:))
end do
end program
```

The file "fort.111" contains the orthogonal matrix whose columns are the orthonormal vectors.

0.182574183	-4.71208207E-02	-0.100089625	-0.976948500
0.365148365	0.930636227	-4.16827016E-03	2.37796791E-02
0.547722578	-0.212043703	0.808797836	2.97245998E-02
0.730296731	-0.294505149	-0.579491854	0.210053831

3.12.2 For complex vectors

```
subroutine GSOMC(n,k,wi,vi)
```

Parameters

In/Out	Argument	Description
[in]	n	n (integer) is the number of rows in the matrix wi
[in]	k	k (integer) is the number of columns in the matrix wi
[in]	wi	wi is a complex*16 array of dimension $(0:n-1,0:k-1)$, whose columns are the linearly independent non-orthonormal set of vectors

In/Out	Argument	Description
[out]	vi	vi is a complex*16 array of dimension (0:n−1,0:k−1) which stores the k vectors as columns of dimension n which is the orthonormalized set of vectors

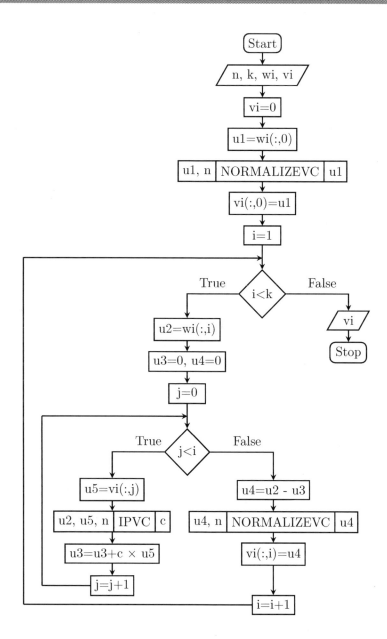

FIGURE 3.32: Flowchart of GSOMC

Implementation

```
subroutine GSOMC(n,k,wi,vi)
implicit none
integer::n,k,i,j
complex*16::wi(0:n-1,0:k-1),vi(0:n-1,0:k-1),c
complex*16,dimension(0:n-1)::u1,u2,u3,u4,u5
vi=0.0d0
u1=wi(:,0)
call NORMALIZEVC(u1,n)
vi(:,0)=u1
do i=1,k-1
   u2=wi(:,i)
   u3=0.0d0
   u4=0.0d0
   do j=0,i-1
      u5=vi(:,j)
      call IPVC(u5,u2,n,c)
      u3=u3+c*u5
   end do
   u4=u2-u3
   call NORMALIZEVC(u4,n)
   vi(:,i)=u4
end do
end subroutine
```

Example

In this example we are doing the Gram-Schmidt orthogonalization of the matrix A whose columns are the linearly independent non-orthogonal vectors.

$$A = \begin{pmatrix} 1+i & 2+i \\ 2-i & 4 \end{pmatrix}. \tag{3.31}$$

```
program example
implicit none
integer::i
complex*16::A(0:1,0:1),m1(0:1,0:1),m2(0:1,0:1)
A(0,0)=dcmplx(1,1)
A(0,1)=dcmplx(2,1)
A(1,0)=dcmplx(2,-1)
A(1,1)=dcmplx(4,0)
call GSOMC(2,2,A,m1)
m2=matmul(conjg(transpose(m1)),m1)
do i=0,1
write(111,*)abs(m2(i,:))
end do
end program
```

The file "fort.111" contains the identity matrix which shows that, $m1^\dagger m1 = \mathbb{I}$.

0.99999999999999978	5.5648107147161955E-016
5.8614635538184244E-016	1.0000000000000000

3.13 Singular value decomposition

Any $m \times n$ real or a complex matrix A can be decomposed as follows,

$$A = U\Sigma V^{\dagger}, \tag{3.32}$$

where U is an $m \times m$ real or complex unitary matrix Σ is an $m \times n$ rectangular diagonal matrix with non-negative real numbers on the main diagonal called the singular values of A and V is an $n \times n$ real or complex unitary matrix [11, 12, 27]. Note also that the square of the singular values will be the eigenvalues of the matrix $A^{\dagger}A$. Alternatively, in terms of the matrix elements of the respective matrices in Eq. [3.32], we can write it as

$$a_{ij} = \sum_{k=1}^{n} u_{ik} s_k v_{jk}, \tag{3.33}$$

here, s_k's are the singular values contained in the matrix Σ, actually the sum goes from 1 to the rank of matrix A, the singular values are normally arranged in the decreasing order and if $m < n$, there will be at most m non-zero singular values. The Singular value decomposition manifests as what is popularly called as the Schmidt decomposition [11] in quantum information theory. The singular value decomposition is a very useful tool when we deal with bipartite splits of quantum systems which are described by pure state vectors. Other uses of SVD includes the calculation of inverse of matrices, solution of homogneous linear systems of equations and so on.

3.13.1 For a real matrix

subroutine SVDMR(A,M,N,U,Si,VT)

Parameters

In/Out	Argument	Description
[in]	A	A is a real*8 array of dimension (0:M−1,0:N−1)
[in]	M	M (integer) is the total number of rows in matrix A
[in]	N	N (integer) is the total number of columns in matrix A
[out]	U	U is a real*8 array of dimension (0:M−1,0:M−1) which is a unitary matrix
[out]	Si	Si is a real*8 array of dimension (0:M−1,0:N−1) which stores the singular values along the main diagonal

In/Out	Argument	Description
[out]	VT	VT is a real*8 array of dimension (0:N−1,0:N−1) which is also a unitary matrix.

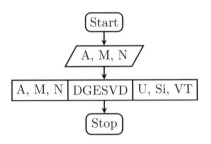

FIGURE 3.33: Flowchart of SVDMR

Implementation

```
subroutine SVDMR(A,M,N,U,Si,VT)
integer::M,N
integer::LWORK
real*8::A(0:m-1,0:n-1),S(0:min(M,N)-1),U(0:M-1,0:M-1)
real*8::VT(0:N-1,0:N-1),B(0:m-1,0:n-1),Si(0:m-1,0:n-1)
character::JOBU, JOBVT
real*8,allocatable,dimension(:)::WORK
integer::INFO
JOBU='A'
JOBVT='A'
LWORK=MAX(3*MIN(M,N)+MAX(M,N),5*MIN(M,N))
allocate(WORK(0:LWORK-1))
call DGESVD(JOBU, JOBVT, M, N, A, M, S, &
            U, M, VT, N,WORK, LWORK, INFO)
Si=0.0d0
do i=0,min(M,N)
Si(i,i)=S(i)
end do
deallocate(WORK)
end subroutine
```

Example

In this example we are doing the singular value decomposition of the matrix given below:

$$A = \begin{pmatrix} 1 & 2 & 3 & 4 \\ 2 & 33 & 4 & 5 \\ 3 & 4 & 53 & 6 \\ 4 & 5 & 6 & 73 \end{pmatrix}. \qquad (3.34)$$

```
program example
implicit none
integer::i,j
real*8::m1(0:3,0:3),m2(0:3,0:3),U(0:3,0:3)
real*8::si(0:3,0:3),VT(0:3,0:3)
m1(0,:)=(/1,2,3,4/)
m1(1,:)=(/2,33,4,5/)
m1(2,:)=(/3,4,53,6/)
m1(3,:)=(/4,5,6,73/)
call SVDMR(m1,4,4,U,si,VT)
write(111,*)'matrix U'
do i=0,3
    do j=0,3
        write(111,*)U(i,j)
    end do
end do
write(111,*)'Singular values'
do j=0,3
    write(111,*)si(j,j)
end do
write(111,*)'matrix VT'
do i=0,3
    do j=0,3
        write(111,*)VT(i,j)
    end do
end do
end program
```

The file "fort.111" contains the entries of matrices U, Si and VT.

```
matrix U
 -6.5690562524116231E-002
  3.6908306077149254E-002
  3.4875171613176575E-002
-0.99654716363187357
-0.14003691010553040
  0.12486168958242540
  0.98105916630265000
  4.8188530754389856E-002
-0.28228875264028153
  0.94419885758245625
-0.16281623030684297
  4.7879563714039543E-002
```

```
-0.94677741520177483
-0.30254892622040697
 -9.8982370991857485E-002
  4.7740585873597641E-002
Singular values
  75.806023119911032
  51.723656705276113
  31.902791523759920
  0.56752865105289452
matrix VT
 -6.5690562524116314E-002
-0.14003691010553057
-0.28228875264028153
-0.94677741520177516
  3.6908306077149587E-002
  0.12486168958242527
  0.94419885758245636
-0.30254892622040663
  3.4875171613176464E-002
  0.98105916630265011
-0.16281623030684295
 -9.8982370991857527E-002
-0.99654716363187357
  4.8188530754389683E-002
  4.7879563714039751E-002
  4.7740585873597613E-002
```

3.13.2 For a complex matrix

subroutine SVDMC(A,M,N,U,Si,VT)

Parameters

In/Out	Argument	Description
[in]	A	A is a complex*16 array of dimennsion (0:M−1,0:N−1)
[in]	M	M (integer) is the total number of rows in matrix A
[in]	N	N (integer) is the total number of columns in matrix A
[out]	U	U is a complex*16 array of dimension (0:M−1,0:M−1) which is a unitary matrix
[out]	Si	Si is a real*8 array of dimension (0:M−1,0:N−1) which stores the singular values along the main diagonal
[out]	VT	VT is a complex*16 array of dimension (0:N−1,0:N−1) which is also a unitary matrix

Implementation

```fortran
subroutine SVDMC(A,M,N,U,Si,VT)
integer::m,n
integer::LWORK
real*8::S(0:min(M,N)-1),Si(0:m-1,0:n-1),RWORK(0:5*min(M,N)-1)
complex*16::A(0:m-1,0:n-1),U(0:M-1,0:M-1),VT(0:N-1,0:N-1)
complex*16::B(0:m-1,0:n-1)
complex*16,allocatable,dimension(:)::WORK
character::JOBU, JOBVT
integer::INFO
JOBU='A'
JOBVT='A'
LWORK=MAX(1,2*MIN(M,N)+MAX(M,N))
allocate(WORK(0:LWORK-1))
call ZGESVD(JOBU, JOBVT, M, N, A, M, S, &
                 U, M, VT, N,WORK, LWORK, RWORK, INFO)
Si=0.0d0
do i=0,min(M,N)
Si(i,i)=S(i)
end do
deallocate(WORK)
end subroutine
```

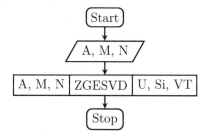

FIGURE 3.34: Flowchart of SVDMC

Example

In this example we are doing the singular value decomposition of the matrix given below:

$$A = \begin{pmatrix} -i & -50 \\ 537 & 180i \end{pmatrix}. \tag{3.35}$$

```fortran
program example
implicit none
integer::i,j
complex*16::A(0:1,0:1),B(0:1,0:1),U(0:1,0:1),VT(0:1,0:1)
real*8::si(0:1,0:1)
A(0,0)=dcmplx(0,-1)
A(0,1)=dcmplx(-50,0)
```

```
A(1,0)=dcmplx(537,0)
A(1,1)=dcmplx(0,180)
call SVDMC(A,2,2,U,Si,VT)
write(111,*)'matrix U'
do i=0,1
    do j=0,1
        write(111,*)U(i,j)
    end do
end do
write(111,*)'Singular values'
do j=0,1
    write(111,*)si(j,j)
end do
write(111,*)'matrix VT'
do i=0,1
    do j=0,1
        write(111,*)VT(i,j)
    end do
end do
end program
```

The file "fort.111" contains entries of matrices U, Si and VT.

```
matrix U
  (2.75077111476836204E-017,-2.65626467434847413E-002)
      (9.67367838679354916E-016,-0.99964715064765741)
      (-0.99964715064765741,9.25038684386347418E-018)
  (2.65626467434847413E-002,3.32938496061536364E-016)
Singular values
  566.56321715039849
  47.708709605171350
matrix VT
          (-0.94743876941194982,0.0000000000000000)
          (-0.0000000000000000,-0.31993714728860451)
          (0.31993714728860451,-0.0000000000000000)
          (-0.0000000000000000,-0.94743876941194982)
```

4

Tools of Quantum Information Theory

In this chapter, we have given the recipes to construct the most important quantum gates [11,28–30] which are used in quantum computing, however, a suitable combination of these gates can be used to accomplish a specific quantum computing task at hand. Given these gates, they have to act on something, those are the quantum states. We have given how to construct many interesting states which are used in quantum computing. These are the prototypical states in their respective Hilbert spaces. Toward the end we have given certain quantities like fidelity, trace distance, entropy, as these are important in studying the closeness between two pure states and related to the entanglement content in them. Last but not least, there are recipes dealing with single qubit quantum measurements in the spin basis and expectation values of observables which help us to have a direct mapping between the Hilbert space which resides in the mind of the mathematician to the laboratory which resides in front of us in reality. The following pure states and density matrices will be mostly used to demonstrate the recipes in this chapter.

Real states used for demonstration are,

$$|\psi_1\rangle = \frac{1}{\sqrt{2}}(|00\rangle + |11\rangle), \tag{4.1}$$

$$|\psi_2\rangle = \frac{1}{\sqrt{c_2}} \sum_{x=1}^{2^N} x|x\rangle = \frac{1}{\sqrt{c_2}} \begin{pmatrix} 1 \\ 2 \\ 3 \\ \vdots \\ 2^N \end{pmatrix}, \tag{4.2}$$

where c_2 is the normalization constant given by,

$$c_2 = \sum_{x=1}^{2^N} x^2. \tag{4.3}$$

Complex states used for demonstration are,

$$|\psi_3\rangle = \frac{1}{\sqrt{c_3}} \sum_{x=1}^{2^N} (x - 1 + ix)|x\rangle = \frac{1}{\sqrt{c_3}} \begin{pmatrix} i \\ 1 + 2i \\ 2 + 3i \\ \vdots \\ 2^N - 1 + 2^N i \end{pmatrix}, \tag{4.4}$$

where c_3 is the normalization constant given by,

$$c_3 = \sum_{x=1}^{2^N}((x-1)^2 + x^2),\qquad(4.5)$$

$$|\psi_4\rangle = \frac{1}{\sqrt{c_4}}\sum_{x=1}^{2^N}(x-1+i(x+1))|j\rangle = \frac{1}{\sqrt{c_4}}\begin{pmatrix} 2i \\ 1+3i \\ 2+4i \\ \vdots \\ 2^N-1+(2^N+1)i \end{pmatrix},\qquad(4.6)$$

where c_4 is the normalization constant given by,

$$c_4 = \sum_{x=1}^{2^N}((x-1)^2 + (x+1)^2).\qquad(4.7)$$

Complex density matrices used for demonstration are,

$$\rho_1 = \begin{pmatrix} 0.4 & -0.2-0.1i \\ -0.2+0.1i & 0.6 \end{pmatrix},\qquad(4.8)$$

$$\rho_2 = \begin{pmatrix} 0.4 & -0.1-0.4i \\ -0.1+0.4i & 0.6 \end{pmatrix}.\qquad(4.9)$$

4.1 Frequently used quantum gates

Quantum gates are important elements in a quantum computing algorithm. Quantum logic gates manifest themselves as unitary matrices and they operate on a quantum state to produce a desired output depending on what operation they do. Gates are classified as one qubit, two-qubit and multi-qubit depending on the state on which they act. Just as how in Boolean algebra we have truth tables for the operations, similarly we have them in the case of quantum gates too. We will be giving the numerical recipes to construct such gates, however, one can build up complex quantum circuits using these gates as building blocks.

4.1.1 The Pauli-X

The Pauli-X gate acts on a single-qubit state. This is the quantum gate equivalent to the NOT gate. It is sometimes also known as a bit flip operator. It maps $|0\rangle$ to $|1\rangle$ and $|1\rangle$ to $|0\rangle$. A generalized version of writing the Pauli-X acting on a qubit will be $X|a\rangle = |1-a\rangle$, where a is either 1 or 0. The matrix representation of Pauli-X is given below:

$$X = \begin{pmatrix} 0 & 1 \\ 1 & 0 \end{pmatrix}.$$

Circuit symbol

Or,

Truth table

Input	Output
$|0\rangle$	$|1\rangle$
$|1\rangle$	$|0\rangle$

---

subroutine PAULIX(sigmax)

Parameters

In/Out	Argument	Description
[out]	sigmax	sigmax is a real*8 array of dimension (0:1,0:1) which is the Pauli matrix X

Implementation

```
subroutine PAULIX(sigmax)
real*8::sigmax(0:1,0:1)
sigmax(0,:)=(/0,1/)
sigmax(1,:)=(/1,0/)
end subroutine
```

Example

```
program example
implicit none
real*8::sigmax(0:1,0:1)
integer::j
call PAULIX(sigmax)
do j=0,1
   write(111,*)sigmax(j,:)
end do
end program
```

The file "fort.111" contains the Pauli matrix X.

```
      0              1
      1              0
```

4.1.2   The Pauli-Y

The Pauli-Y gate acts on a single-qubit state. It maps $|0\rangle$ to $i|1\rangle$ and $|1\rangle$ to $-i|0\rangle$. It is clear that, the Pauli-Y gate is similar to the Pauli-X gate but with an additional phase of $\pm i$. A generalized version of writing the Pauli-Y acting on a qubit will be $Y|a\rangle = (-1)^a i|1 - a\rangle$ where a is 0 or 1. The matrix representation of Pauli-Y is given below:

$$Y = \begin{pmatrix} 0 & -i \\ i & 0 \end{pmatrix}.$$

Circuit symbol

Truth table

Input	Output
$|0\rangle$	$i|1\rangle$
$|1\rangle$	$-i|0\rangle$

---

subroutine PAULIY(sigmay)

Parameters

In/Out	Argument	Description
[out]	sigmay	sigmay is a complex*16 array of dimension (0:1,0:1) which is the Pauli matrix Y

Implementation

```
subroutine PAULIY(sigmay)
complex*16::sigmay(0:1,0:1)
sigmay(0,0)=dcmplx(0,0)
sigmay(0,1)=dcmplx(0,-1)
sigmay(1,0)=dcmplx(0,1)
sigmay(1,1)=dcmplx(0,0)
end subroutine
```

Example

```
program example
implicit none
complex*16::sigmay(0:1,0:1)
integer::j,i
call PAULIY(sigmay)
```

```
do i=0,1
    do j=0,1
        write(111,*)sigmay(i,j)
    end do
end do
end program
```

The file "fort.111" contains entries of the Pauli matrix Y.

```
(0.0000000000000000,0.0000000000000000)
(0.0000000000000000,-1.0000000000000000)
(0.0000000000000000,1.0000000000000000)
(0.0000000000000000,0.0000000000000000)
```

4.1.3 The Pauli-Z

The Pauli-Z gate acts on a single-qubit state. It leaves $|0\rangle$ unchanged and maps $|1\rangle$ to $-|1\rangle$, the reason being, the Pauli-Z gate is already in the diagonal basis. A generalized version of writing the Pauli-Z acting on a qubit will be $Z|a\rangle = (1 - 2a)|a\rangle$, where a is 0 or 1. The matrix representation of Pauli-Z is given below:

$$Z = \begin{pmatrix} 1 & 0 \\ 0 & -1 \end{pmatrix}.$$

Circuit symbol

Truth table

Input	Output
$|0\rangle$	$|0\rangle$
$|1\rangle$	$-|1\rangle$

```
subroutine PAULIZ(sigmaz)
```

Parameters

In/Out	Argument	Description
[out]	sigmaz	sigmaz is a real*8 array of dimension (0:1,0:1) which is the Pauli matrix Z

Implementation

```
subroutine PAULIZ(sigmaz)
real*8::sigmaz(0:1,0:1)
sigmaz(0,:)=(/1,0/)
sigmaz(1,:)=(/0,-1/)
end subroutine
```

Example

```
program example
implicit none
real*8::sigmaz(0:1,0:1)
integer::j
call PAULIZ(sigmaz)
do j=0,1
    write(111,*)int(sigmaz(j,:))
end do
end program
```

The file "fort.111" contains the Pauli matrix Z.

1	0
0	-1

4.1.4   The Hadamard gate

The Hadamard gate acts on a single-qubit state. It maps $|0\rangle$ to $\frac{|0\rangle+|1\rangle}{\sqrt{2}}$ and $|1\rangle$ to $\frac{|0\rangle-|1\rangle}{\sqrt{2}}$. The Hadamard gate is often used to interconvert between the Pauli-X and the Pauli-Z eigenstates. It is to be noted that the columns of the Hadamard matrix are the eigenvectors of Pauli-X when represented in the computational basis. The matrix representation of the Hadamard gate is given below:

$$H = \frac{1}{\sqrt{2}} \begin{pmatrix} 1 & 1 \\ 1 & -1 \end{pmatrix}.$$

Circuit symbol

$$-\boxed{H}-$$

Truth table

Input	Output
$|0\rangle$	$\frac{|0\rangle+|1\rangle}{\sqrt{2}}$
$|1\rangle$	$\frac{|0\rangle-|1\rangle}{\sqrt{2}}$

```
subroutine HADAMARDG(H)
```

Parameters

In/Out	Argument	Description
[out]	H	H is a real*8 array of dimension (0:1,0:1) which is the Hadamard matrix

Implementation

```
subroutine HADAMARDG(H)
real*8::H(0:1,0:1),ele
ele=1/dsqrt(2.0d0)
H(0,:)=(/ele,ele/)
H(1,:)=(/ele,-ele/)
end subroutine
```

Example

```
program example
implicit none
real*8::H(0:1,0:1)
integer::j
call HADAMARDG(H)
do j=0,1
    write(111,*)H(j,:)
end do
end program
```

The file "fort.111" contains the Hadamard matrix.

0.70710678118654746	0.70710678118654746
0.70710678118654746	-0.70710678118654746

4.1.5 The phase gate

The phase gate acts on a single-qubit state. It leaves $|0\rangle$ unchanged and maps $|1\rangle$ to $i|1\rangle$ where is a phase factor. The matrix representation of the Phase gate is given below:

$$S = \begin{pmatrix} 1 & 0 \\ 0 & i \end{pmatrix}.$$

Circuit symbol

Truth table

Input	Output
$\vert 0\rangle$	$\vert 0\rangle$
$\vert 1\rangle$	$i\vert 1\rangle$

---
subroutine PHASEG(S)

Parameters

In/Out	Argument	Description
[out]	S	S is a complex*16 array of dimension (0:1,0:1) which is the phase gate

Implementation

```
subroutine PHASEG(S)
complex*16::S(0:1,0:1)
S(0,0)=dcmplx(1,0)
S(0,1)=dcmplx(0,0)
S(1,0)=dcmplx(0,0)
S(1,1)=dcmplx(0,1)
end subroutine
```

Example

```
program example
implicit none
complex*16::S(0:1,0:1)
integer::j,i
call PHASEG(S)
do i=0,1
   do j=0,1
      write(111,*)S(i,j)
   end do
end do
end program
```

The file "fort.111" contains entries of the phase gate.

```
(1.0000000000000000,0.0000000000000000)
(0.0000000000000000,0.0000000000000000)
(0.0000000000000000,0.0000000000000000)
(0.0000000000000000,1.0000000000000000)
```

4.1.6 The rotation gate

The rotation gate acts on a single-qubit state. The matrix representation of the rotation gate is given below:

$$R_k = \begin{pmatrix} 1 & 0 \\ 0 & e^{\frac{2\pi i}{2^k}} \end{pmatrix}.$$

Where k is a positive integer.

Circuit symbol

Truth table

Input	Output
$|0\rangle$	$|0\rangle$
$|1\rangle$	$e^{\frac{2\pi i}{2^k}}|1\rangle$

Note that, when $k = 2$, we get the phase gate S, meaning that $S = R_2$.

```
subroutine ROTG(R,k)
```

Parameters

In/Out	Argument	Description
[in]	k	k is a positive integer
[out]	R	R is a complex*16 array of dimension (0:1,0:1) which is the rotation gate

Implementation

```
subroutine ROTG(R,k)
integer::k
real*8::pi
complex*16::R(0:1,0:1),i1,ele
pi=4*atan(1.0d0)
i1=dcmplx(0,1)
ele=exp((2*pi*i1)/(2**k))
R(0,0)=dcmplx(1,0)
R(0,1)=dcmplx(0,0)
R(1,0)=dcmplx(0,0)
R(1,1)=ele
end subroutine
```

Example

```
program example
implicit none
complex*16::R(0:1,0:1)
integer::j,i
call ROTG(R,2)
do i=0,1
    do j=0,1
        write(111,*)R(i,j)
    end do
end do
end program
```

The file "fort.111" contains entries of the rotation gate.

$$(1.0000000000000000,0.0000000000000000)$$
$$(0.0000000000000000,0.0000000000000000)$$
$$(0.0000000000000000,0.0000000000000000)$$
$$(6.12323399573676604E\text{-}017,1.0000000000000000)$$

4.1.7   The controlled NOT gate (CX gate)

The controlled NOT gate acts on a two-qubit state. The controlled-NOT X gate performs a Pauli-X operation on the second qubit (called target bit) only when the first qubit (called control bit) is $|1\rangle$, otherwise leaves it unchanged. In the standard two-qubit orthonormal basis $\{|00\rangle, |01\rangle, |10\rangle, |11\rangle\}$, the matrix representation of the controlled NOT gate is given below:

$$CX = \begin{pmatrix} 1 & 0 & 0 & 0 \\ 0 & 1 & 0 & 0 \\ 0 & 0 & 0 & 1 \\ 0 & 0 & 1 & 0 \end{pmatrix}.$$

Circuit symbol

Or,

Truth table

Input	Output
$|00\rangle$	$|00\rangle$
$|01\rangle$	$|01\rangle$
$|10\rangle$	$|11\rangle$
$|11\rangle$	$|10\rangle$

---

subroutine CONOTG(A)

Parameters

In/Out	Argument	Description
[out]	A	A is a real*8 array of dimension (0:3,0:3) which is the controlled NOT gate

Implementation

```
subroutine CONOTG(A)
real*8::A(0:3,0:3)
A(0,:)=(/1,0,0,0/)
A(1,:)=(/0,1,0,0/)
A(2,:)=(/0,0,0,1/)
A(3,:)=(/0,0,1,0/)
end subroutine
```

Example

In this example we are illustrating the fourth operation of the truth table which is, $CX|11\rangle = |10\rangle$.

```
program example
implicit none
real*8::A(0:3,0:3),state(0:3)
integer::j
call CONOTG(A)
state=0.0d0
state(3)=1
state=matmul(A,state)
do j=0,3
    write(111,*)int(state(j))
end do
end program
```

The file "fort.111" contains the output of the fourth operation of the truth table.

0
0
1
0

4.1.8   The controlled Z gate (CZ gate)

The controlled Z gate acts on a two-qubit state. The controlled NOT Z gate performs a Pauli-Z operation on the second qubit (target bit) only when the first qubit (control bit) is $|1\rangle$. The matrix representation of the controlled Z gate is given below:

$$CZ = \begin{pmatrix} 1 & 0 & 0 & 0 \\ 0 & 1 & 0 & 0 \\ 0 & 0 & 1 & 0 \\ 0 & 0 & 0 & -1 \end{pmatrix}.$$

Circuit symbol

Truth table

Input	Output
$|00\rangle$	$|00\rangle$
$|01\rangle$	$|01\rangle$
$|10\rangle$	$|10\rangle$
$|11\rangle$	$-|11\rangle$

subroutine COZG(A)

Parameters

In/Out	Argument	Description
[out]	A	A is a real*8 array of dimension (0:3,0:3) which is the controlled Z gate.

Implementation

subroutine COZG(A)

```
real*8::A(0:3,0:3)
A(0,:)=(/1,0,0,0/)
A(1,:)=(/0,1,0,0/)
A(2,:)=(/0,0,1,0/)
A(3,:)=(/0,0,0,-1/)
end subroutine
```

Example

In this example we are illustrating the fourth operation of the truth table which is, $CZ|11\rangle = -|11\rangle$.

```
program example
implicit none
real*8::A(0:3,0:3),state(0:3)
integer::j
call COZG(A)
state=0.0d0
state(3)=1
state=matmul(A,state)
do j=0,3
    write(111,*)int(state(j))
end do
end program
```

The file "fort.111" contains the output of the fourth operation of the truth table.

```
     0
     0
     0
    -1
```

4.1.9   The SWAP gate

The SWAP gate swaps or permutes the bits of a two-qubit state. For example, the SWAP gate operating on $|10\rangle$ will result in $|01\rangle$. The matrix representation of the SWAP gate is given below:

$$SWAP = \begin{pmatrix} 1 & 0 & 0 & 0 \\ 0 & 0 & 1 & 0 \\ 0 & 1 & 0 & 0 \\ 0 & 0 & 0 & 1 \end{pmatrix}.$$

Circuit symbol

Truth table

Input	Output
$|00\rangle$	$|00\rangle$
$|01\rangle$	$|10\rangle$
$|10\rangle$	$|01\rangle$
$|11\rangle$	$|11\rangle$

```
subroutine SWAPG(A)
```

Parameters

In/Out	Argument	Description
[out]	A	A is a real*8 array of dimension (0:3,0:3) which is the SWAP gate

Implementation

```
subroutine SWAPG(A)
real*8::A(0:3,0:3)
A(0,:)=(/1,0,0,0/)
A(1,:)=(/0,0,1,0/)
A(2,:)=(/0,1,0,0/)
A(3,:)=(/0,0,0,1/)
end subroutine
```

Example

In this example we are illustrating the third operation of the truth table which is, $\text{SWAP}|10\rangle = |01\rangle$.

```
program example
implicit none
real*8::A(0:3,0:3),state(0:3)
integer::j
call SWAPG(A)
state=0.0d0
state(2)=1
state=matmul(A,state)
do j=0,3
    write(111,*)int(state(j))
end do
end program
```

The file "fort.111" contains the output of the third operation of the truth table.

$$\begin{matrix} 0 \\ 1 \\ 0 \\ 0 \end{matrix}$$

4.1.10 The Toffoli gate (CCX, CCNOT, TOFF)

The Toffoli gate [31, 32], named after Tommaso Toffoli; is also called the CCNOT gate. The Toffoli gate acts on a three-qubit state. When the first two qubits are $|1\rangle$ then only it performs a Pauli-X operation on the third qubit. The matrix representation of the Toffoli gate is given below:

$$CCX = \begin{pmatrix} 1 & 0 & 0 & 0 & 0 & 0 & 0 & 0 \\ 0 & 1 & 0 & 0 & 0 & 0 & 0 & 0 \\ 0 & 0 & 1 & 0 & 0 & 0 & 0 & 0 \\ 0 & 0 & 0 & 1 & 0 & 0 & 0 & 0 \\ 0 & 0 & 0 & 0 & 1 & 0 & 0 & 0 \\ 0 & 0 & 0 & 0 & 0 & 1 & 0 & 0 \\ 0 & 0 & 0 & 0 & 0 & 0 & 0 & 1 \\ 0 & 0 & 0 & 0 & 0 & 0 & 1 & 0 \end{pmatrix}.$$

Circuit symbol

Truth table

Input	Output
$|000\rangle$	$|000\rangle$
$|001\rangle$	$|001\rangle$
$|010\rangle$	$|010\rangle$
$|011\rangle$	$|011\rangle$
$|100\rangle$	$|100\rangle$
$|101\rangle$	$|101\rangle$
$|110\rangle$	$|111\rangle$
$|111\rangle$	$|110\rangle$

```
subroutine TOFFOLIG(A)
```

Parameters

In/Out	Argument	Description
[out]	A	A is a real*8 array of dimension (0:7,0:7) which is the Toffoli gate

Implementation

```
subroutine TOFFOLIG(A)
real*8::A(0:7,0:7)
A(0,:)=(/1,0,0,0,0,0,0,0/)
A(1,:)=(/0,1,0,0,0,0,0,0/)
A(2,:)=(/0,0,1,0,0,0,0,0/)
A(3,:)=(/0,0,0,1,0,0,0,0/)
A(4,:)=(/0,0,0,0,1,0,0,0/)
A(5,:)=(/0,0,0,0,0,1,0,0/)
A(6,:)=(/0,0,0,0,0,0,0,1/)
A(7,:)=(/0,0,0,0,0,0,1,0/)
end subroutine
```

Example

In this example we are illustrating the seventh operation of the truth table which is, $CCX|110\rangle = |111\rangle$.

```
program example
implicit none
real*8::A(0:7,0:7),state(0:7)
integer::j
call TOFFOLIG(A)
state=0.0d0
state(6)=1
state=matmul(A,state)
do j=0,7
    write(111,*)int(state(j))
end do
end program
```

The file "fort.111" contains the output of the seventh operation of the truth table.

```
            0
            0
            0
            0
            0
            0
```

0

1

4.1.11 The Fredkin gate (CSWAP)

The Fredkin gate, [33] (also CSWAP or cS gate), named after Edward Fredkin is a three qubit gate that performs a controlled swap operation. It simply means that, when the control bit is 1, it swaps the second and the third bit. An interesting feature of the Fredkin gate is that it can be used as an universal gate and also to do reversible computing. The matrix representation of the Fredkin gate is given below:

$$X = \begin{pmatrix} 1 & 0 & 0 & 0 & 0 & 0 & 0 & 0 \\ 0 & 1 & 0 & 0 & 0 & 0 & 0 & 0 \\ 0 & 0 & 1 & 0 & 0 & 0 & 0 & 0 \\ 0 & 0 & 0 & 1 & 0 & 0 & 0 & 0 \\ 0 & 0 & 0 & 0 & 1 & 0 & 0 & 0 \\ 0 & 0 & 0 & 0 & 0 & 0 & 1 & 0 \\ 0 & 0 & 0 & 0 & 0 & 1 & 0 & 0 \\ 0 & 0 & 0 & 0 & 0 & 0 & 0 & 1 \end{pmatrix}.$$

Circuit symbol

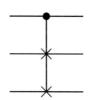

Truth table

Input	Output
$|000\rangle$	$|000\rangle$
$|001\rangle$	$|001\rangle$
$|010\rangle$	$|010\rangle$
$|011\rangle$	$|011\rangle$
$|100\rangle$	$|100\rangle$
$|101\rangle$	$|110\rangle$
$|110\rangle$	$|101\rangle$
$|111\rangle$	$|111\rangle$

subroutine FREDKING(A)

Parameters

In/Out	Argument	Description
[out]	A	A is a real*8 array of dimension (0:7,0:7) which is the Fredkin gate

Implementation

```
subroutine FREDKING(A)
real*8::A(0:7,0:7)
A(0,:)=(/1,0,0,0,0,0,0,0/)
A(1,:)=(/0,1,0,0,0,0,0,0/)
A(2,:)=(/0,0,1,0,0,0,0,0/)
A(3,:)=(/0,0,0,1,0,0,0,0/)
A(4,:)=(/0,0,0,0,1,0,0,0/)
A(5,:)=(/0,0,0,0,0,0,1,0/)
A(6,:)=(/0,0,0,0,0,1,0,0/)
A(7,:)=(/0,0,0,0,0,0,0,1/)
end subroutine
```

Example

In this example we are illustrating the seventh operation of the truth table which is, $\text{CSWAP}|110\rangle = |101\rangle$.

```
program example
implicit none
real*8::A(0:7,0:7),state(0:7)
integer::j
call FREDKING(A)
state=0.0d0
state(6)=1
state=matmul(A,state)
do j=0,7
    write(111,*)int(state(j))
end do
end program
```

The file "fort.111" contains the output of the seventh operation of the truth table.

```
           0
           0
           0
           0
           0
           1
```

      0

      0

4.2   The N-qubit Hadamard gate

When we apply the Hadamard gate to each qubit in a N bit string, then the corresponding operator is the N-qubit Hadamard gate [11]. It is a very important gate and in fact a very important quantum computing element. It is used extensively in important operations such as the Deutsch–Jozsa algorithm [34], quantum Fourier transform [35] and many more. The mathematical expression for the N-qubit Hadamard gate is given by,

$$H^{\otimes n} = \frac{1}{\sqrt{2^N}} \sum_{xy} (-1)^{x.y} |x\rangle\langle y|, \tag{4.10}$$

where $x, y = 0$ or 1. Also note the dot between x and y in the power of -1.

subroutine NQBHADAMARDG(s,A)

Parameters

In/Out	Argument	Description
[in]	s	s (integer) is the total number of qubits
[out]	A	A is a real*8 array of dimension (0:2**s−1,0:2**s−1) which is the s qubit Hadamard matrix

Implementation

```
subroutine NQBHADAMARDG(s,A)
implicit none
integer::i,j,s,t1(0:s-1),t2(0:s-1),k,xy
real*8::A(0:2**s-1,0:2**s-1)
A=0.0d0
do i=0,2**s-1
   do j=0,2**s-1
       call DTOB(i,t1,s)
       call DTOB(j,t2,s)
       xy=0
       do k=0,s-1
           xy=xy+t1(k)*t2(k)
       end do
       A(j,i)=(1/dsqrt(dfloat(2**s)))*((-1)**(xy))
   end do
end do
end subroutine
```

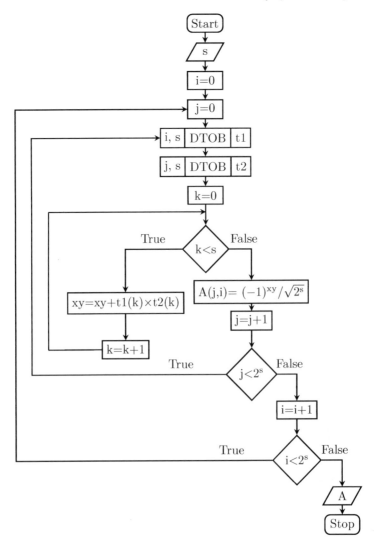

FIGURE 4.1: Flowchart of NQBHADAMARDG

Example

```
program example
implicit none
integer,parameter::s=1
real*8::H(0:2**s-1,0:2**s-1)
integer::j
call NQBHADAMARDG(s,H)
do j=0,2**s-1
    write(111,*)H(j,:)
end do
end program
```

The file "fort.111" contains the two qubit Hadamard gate.

| 0.70710678118654746 | 0.70710678118654746 |
| 0.70710678118654746 | -0.70710678118654746 |

4.3   The N-qubit computational basis (identity matrix)

In quantum mechanics and quantum information theory, the one-qubit state [11, 36] is an important prototypical quantum state from which other higher qubit states can be obtained via the operation of Kronecker product [37]. Any one-qubit state $|\psi\rangle$ is a linear superposition of the orthonormal basis $|0\rangle$ and $|1\rangle$, meaning $|\psi\rangle = a|0\rangle + b|1\rangle$, such that a, b are complex numbers and $|a|^2 + |b|^2 = 1$. The matrix representation of $|0\rangle$ and $|1\rangle$ which are the eigenvectors of the Pauli σ_z matrix or the Pauli-Z gate is as follows,

$$|\uparrow\rangle = |0\rangle = \begin{pmatrix} 1 \\ 0 \end{pmatrix}, \qquad |\downarrow\rangle = |1\rangle = \begin{pmatrix} 0 \\ 1 \end{pmatrix}. \tag{4.11}$$

Here $|0\rangle$ and $|1\rangle$ represent 'up' and 'down' spins, respectively of the spin half particle like an electron or a proton and they are binary elements [11]. The single-qubit states $|0\rangle$ and $|1\rangle$ collectively constitute the computational basis which spans the two-dimensional Hilbert space. We can get the two qubit basis states by taking all possible Kronecker products between $|0\rangle$ and $|1\rangle$ as follows,

$$|00\rangle = \begin{pmatrix} 1 \\ 0 \\ 0 \\ 0 \end{pmatrix}, \quad |01\rangle = \begin{pmatrix} 0 \\ 1 \\ 0 \\ 0 \end{pmatrix}, \quad |10\rangle = \begin{pmatrix} 0 \\ 0 \\ 1 \\ 0 \end{pmatrix}, \quad |11\rangle = \begin{pmatrix} 0 \\ 0 \\ 0 \\ 1 \end{pmatrix}. \tag{4.12}$$

Similarly, we can get the n-qubit states which span the 2^n-dimensional Hilbert space. Note that the n-qubit state will be a column vector of dimension $2^n \times 1$. These states in n dimension are very important from the point of view of various quantum information theoretic studies and they are called the standard basis states.

subroutine NQBCB(n,II)

Parameters

In/Out	Argument	Description
[in]	n	n (integer) is the dimension of II
[out]	II	II is a real*8 array of the dimension (0:n−1,0:n−1) whose columns denote the n-dimensional computational basis

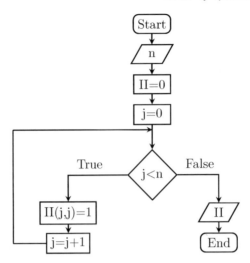

FIGURE 4.2: Flowchart of NQBCB

Implementation

```
subroutine NQBCB(n,II)
implicit none
integer::n,j
real*8::II(0:n-1,0:n-1)
do j=0,n-1,1
    II(j,j)=1.0d0
end do
end subroutine
```

Example

```
program example
implicit none
integer,parameter::s=2
integer::i
real*8::II(0:2**s-1,0:2**s-1)
call NQBCB(2**s,II)
do i=0,2**s-1
    write(111,*)int(II(i,:))
end do
end program
```

The file "fort.111" contains the two qubit computational basis.

1	0	0	0
0	1	0	0
0	0	1	0
0	0	0	1

4.4 The Bell states

The Bell states are four specific maximally entangled two-qubit states [11, 38] as shown below. It is also worth to mention that they contribute another set of orthonormal basis in two-qubit Hilbert space apart from the standard two-qubit basis. Bell states are very important from the point of view of operations such as teleportation [39], superdense coding [40] and to produce a simplified version of the EPR paradox [41].

$$|b_1\rangle = \frac{1}{\sqrt{2}}(|00\rangle + |11\rangle), \tag{4.13}$$

$$|b_2\rangle = \frac{1}{\sqrt{2}}(|00\rangle - |11\rangle), \tag{4.14}$$

$$|b_3\rangle = \frac{1}{\sqrt{2}}(|01\rangle + |10\rangle), \tag{4.15}$$

$$|b_4\rangle = \frac{1}{\sqrt{2}}(|01\rangle - |10\rangle). \tag{4.16}$$

The states $|b_1\rangle$, $|b_2\rangle$, $|b_3\rangle$ and $|b_4\rangle$ together form a maximally entangled two-qubit basis of the four-dimensional Hilbert space called the Bell basis.

```
subroutine BELL(b1,b2,b3,b4)
```

Parameters

In/Out	Argument	Description
[out]	b1, b2, b3, b4	b1, b2, b3 and b4 are real*8 arrays of dimension (0:3) which are the corresponding Bell states

Implementation

```
subroutine BELL(b1,b2,b3,b4)
implicit none
real*8::II(0:3,0:3),b1(0:3),b2(0:3),b3(0:3),b4(0:3)
II(0,:)=(/1,0,0,0/)
II(1,:)=(/0,1,0,0/)
II(2,:)=(/0,0,1,0/)
II(3,:)=(/0,0,0,1/)
b1=(II(:,0)+II(:,3))/dsqrt(2.0d0)
b2=(II(:,0)-II(:,3))/dsqrt(2.0d0)
b3=(II(:,1)+II(:,2))/dsqrt(2.0d0)
b4=(II(:,1)-II(:,2))/dsqrt(2.0d0)
end subroutine
```

Example

In this example we are generating the Bell states b1 and b2.

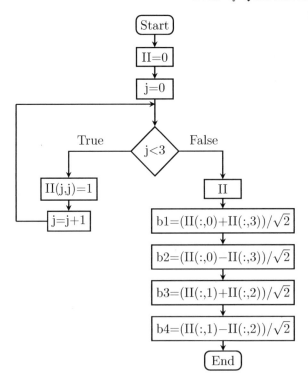

FIGURE 4.3: Flowchart of BELL

```
program example
implicit none
real*8::b1(0:3),b2(0:3),b3(0:3),b4(0:3)
integer::i
call BELL(b1,b2,b3,b4)
do i=0,3
    write(111,*)b1(i),b2(i)
end do
end program
```

The file "fort.111" contains the first two Bell states.

0.70710678118654746	0.70710678118654746
0.0000000000000000	0.0000000000000000
0.0000000000000000	0.0000000000000000
0.70710678118654746	-0.70710678118654746

4.5 The *N*-qubit Greenberger-Horne-Zeilinger (GHZ) state

The GHZ state [42] is an entangled state of three qubits which is,

$$|GHZ\rangle = \frac{|000\rangle + |1111\rangle}{\sqrt{2}}. \tag{4.17}$$

A generalization of it to N qubits ($N \geq 3$) gives us the N-qubit GHZ state which can be defined as

$$|GHZ\rangle = \frac{|0\rangle^{\otimes N} + |1\rangle^{\otimes N}}{\sqrt{2}} = \frac{|00000\cdots 0\rangle_N + |11111\cdots 1\rangle_N}{\sqrt{2}}. \tag{4.18}$$

These states are extensively used in various quantum cryptography protocols and in tasks such as quantum secret sharing. It is in fact quite interesting to note that the N-qubit GHZ state is the ground state of a ferromagnet [43]. Another interesting property of the three-qubit GHZ state ($N = 3$ case) is that, all the three qubits are entangled but no two-qubits are entangled.

```
subroutine NQBGHZ(s,psi)
```

Parameters

In/Out	Argument	Description
[in]	s	s (integer) is the total number of qubits
[out]	psi	psi is a real*8 array of dimension (0:2**s−1) which is the s qubit GHZ state

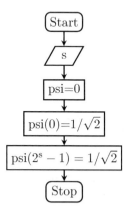

FIGURE 4.4: Flowchart of NQBGHZ

Implementation

```
subroutine NQBGHZ(s,psi)
implicit none
```

```
integer::s
real*8::psi(0:2**s-1)
psi=0.0d0
psi(0)=1.0d0/dsqrt(2.0d0)
psi(2**s-1)=1.0d0/dsqrt(2.0d0)
end subroutine
```

Example

```
program example
implicit none
integer,parameter::s=3
real*8::psi(0:2**s-1)
integer::i
call NQBGHZ(s,psi)
do i=0,2**s-1
   write(111,*)psi(i)
end do
end program
```

The file "fort.111" contains the three qubit GHZ state.

```
0.70710678118654746
0.0000000000000000
0.0000000000000000
0.0000000000000000
0.0000000000000000
0.0000000000000000
0.0000000000000000
0.70710678118654746
```

4.6   The N-qubit W state

The W state [44] is an entangled quantum state of three qubits which can be written as

$$|W\rangle = \frac{1}{\sqrt{3}} \left(|100\rangle + |010\rangle + |001\rangle \right). \tag{4.19}$$

It is very different from the three qubit GHZ state in terms of entanglement properties, that is, all the three qubits are entangled and even the two-qubit subsytems are entangled. It is also interesting to note that the three states making up the W state are such that there is a shift of the qubit $|1\rangle$, either to the right or left with respect to the first state. A natural generalization of the W state to the N qubit case is as follows,

$$|W\rangle = \frac{1}{\sqrt{N}} \left(|100\cdots0\rangle + |010\cdots0\rangle + \cdots + |000\cdots1\rangle \right). \tag{4.20}$$

```
subroutine NQBWSTATE(s,psi)
```

Parameters

In/Out	Argument	Description
[in]	s	s (integer) is the number of qubits
[out]	psi	psi is a real*8 array of dimension (0:2**s−1) which is the s qubit W state

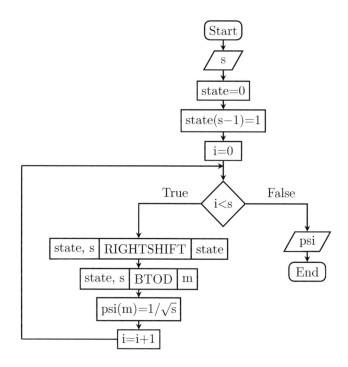

FIGURE 4.5: Flowchart of NQBWSTATE

Implementation

```
subroutine NQBWSTATE(s,psi)
integer::s,m
real*8::psi(0:2**s-1)
integer::state(0:s-1)
psi=0.0d0
state=0
state(s-1)=1
do i=0,s-1
    call RIGHTSHIFT(state,s,1)
    call BTOD(state,m,s)
    psi(m)=1.0/dsqrt(dfloat(s))
end do
end subroutine
```

Example

```
program example
implicit none
integer,parameter::s=3
real*8::psi(0:2**s-1)
integer::i
call NQBWSTATE(s,psi)
do i=0,7
   write(111,*)psi(i)
end do
end program
```

The file "fort.111" contains the three qubit W state.

```
 0.0000000000000000
 0.57735026918962584
 0.57735026918962584
 0.0000000000000000
 0.57735026918962584
 0.0000000000000000
 0.0000000000000000
 0.0000000000000000
```

4.7   The Generalized N-qubit Werner state

The generalized N-qubit Werner [42,45] state can be written in terms of the N-qubit GHZ state and an N-qubit maximally mixed state which are mixed suitably with probabilities p and $1 - p$, respectively,

$$\rho(p) = p|GHZ\rangle\langle GHZ| + \frac{1-p}{2^N}\mathbb{I}, \qquad (4.21)$$

where $\mathbb{I}$ is $2^N \times 2^N$ identity matrix. **Hence forth we will be using $\rho(p)$ as an example of a real density matrix to illustrate the recipes and outputs.**

```
subroutine NQBWERNER(s,x,rho)
```

Parameters

In/Out	Argument	Description
[in]	s	s (integer) is the number of qubits
[in]	x	x (integer) is the mixing probability
[out]	rho	rho is a real*8 array of dimension $(0{:}2{**}\text{s}{-}1,0{:}2{**}\text{s}{-}1)$ which is the s qubit Werner state

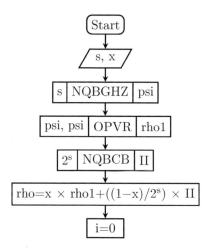

FIGURE 4.6: Flowchart of NQBWERNER

Implementation

```
subroutine NQBWERNER(s,x,rho)
implicit none
integer::s,i,j
real*8,dimension(0:2**s-1,0:2**s-1)::rho,rho1,II
real*8::psi(0:2**s-1),x
call NQBGHZ(s,psi)
call OPVR(psi,psi,2**s,2**s,rho1)
call NQBCB(2**s,II)
rho=x*rho1+((1-x)/2**s)*II
end subroutine
```

Example

```
program example
real*8::rho(0:3,0:3)
integer::i
call NQBWERNER(2,0.5d0,rho)
do i=0,3
    write(111,*)real(rho(i,:))
end do
end program
```

The file "fort.111" contains the two qubit generalized Werner state.

0.375000000	0.00000000	0.00000000	0.250000000
0.00000000	0.125000000	0.00000000	0.00000000
0.00000000	0.00000000	0.125000000	0.00000000
0.250000000	0.00000000	0.00000000	0.375000000

4.8   Shannon entropy

If we have a discrete random variable X, which has outcomes $\{x_i; i = 1, \cdots N\}$ which occur with probabilites $\{p_i(x); i = 1, \cdots N\}$, respectively, then the Shannon entropy [46] for the above discrete set of probabilities is given as,

$$H(p) = -\sum_{i=1}^{N} p_i(x) \log_2 p_i(x). \qquad (4.22)$$

Shannon entropy is one of the most important information theoretic measures widely used. Note that, when $N = 2$, the corresponding entropy is called the binary entropy which is,

$$H_b(p) = -p \log_2 p - (1 - p) \log_2(1 - p). \qquad (4.23)$$

Shannon entropy is useful while computing quantities like mutual information, conditional entropies [47].

subroutine SE(p,sp,n)

Parameters

In/Out	Argument	Description
[in]	n	n (integer) is the total number of probabilities
[in]	p	p is a real*8 array of dimension (0:n−1) which contains probabilities
[out]	sp	sp (real*8) is the Shannon entropy

Implementation

```
subroutine SE(p,sp,n)
implicit none
integer::n,i
real*8::sp,p(0:n-1),sptemp
sp=0
do i=0,n-1,1
    sptemp=-p(i)*dlog(p(i))/dlog(2.0d0)
    sp=sp+sptemp
end do
end subroutine
```

Example

In this example we are finding the Shannon entropy of the probability vector $p = (0.3, 0.2, 0.5)$.

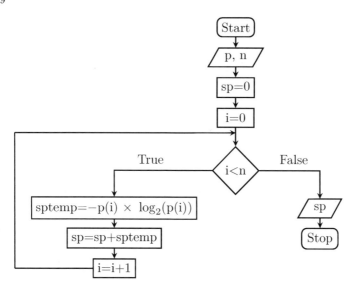

FIGURE 4.7: Flowchart of SE

```
program example
implicit none
real*8::p(0:2),sp
p=(/0.3d0, 0.2d0,0.5d0/)
call SE(p,sp,3)
write(111,*)sp
end program
```

The file "fort.111" contains the required Shannon entropy.

```
1.4854752972273344
```

4.9 Linear entropy

In quantum mechanics, linear entropy is the measure of the impurity of state, it is defined as

$$S_L = 1 - Tr(\rho^2). \tag{4.24}$$

For any state ρ, we know that $Tr(\rho^2) = 1$, if it is pure and $Tr(\rho^2) < 1$ if it is mixed. The linear entropy in this way is a measure of how much a given state deviates from being a pure state.

4.9.1 For a real density matrix

```
subroutine LEDMR(A,n,le)
```

Parameters

In/Out	Argument	Description
[in]	n	n (integer) is the dimension of the matrix A which is ρ
[in/out]	A	A is a real*8 array of dimension (0:n−1,0:n−1), on exit, it will be overwritten
[out]	le	le (real*8) is the linear entropy of ρ

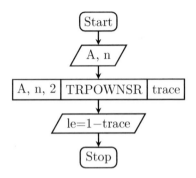

FIGURE 4.8: Flowchart of LEDMR

Implementation

```
subroutine LEDMR(A,n,le)
implicit none
integer::i,n
real*8::A(0:n-1,0:n-1),le
complex*16::trace
call TRPOWNSR(A,n,2.0d0,trace)
le=1-real(trace)
end subroutine
```

Example

In this example we are finding the linear entropy of a two-qubit generalized Werner state $\rho(0)$.

```
program example
implicit none
integer,parameter::s=2
real*8::rho(0:2**s-1,0:2**s-1),le
call NQBWERNER(2,0.0d0,rho)
call LEDMR(rho,2**s,le)
write(111,*)le
end program
```

The file "fort.111" contains the required linear entropy.

0.75000000000000000

4.9.2 For a complex density matrix

subroutine LEDMC(A,n,le)

Parameters

In/Out	Argument	Description
[in]	n	n (integer) is the dimension of the matrix A which is ρ
[in/out]	A	A is a complex*16 array of dimension (0:n−1,0:n−1), on exit, it will be overwritten
[out]	le	le (real*8) is the linear entropy of ρ

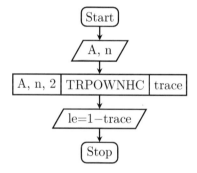

FIGURE 4.9: Flowchart of LEDMC

Implementation

```
subroutine LEDMC(A,n,le)
implicit none
integer::i,n
real*8::le
complex*16::A(0:n-1,0:n-1),trace
call TRPOWNHC(A,n,2.0d0,trace)
le=1-real(trace)
end subroutine
```

Example

In this example we are finding the linear entropy of ρ_1.

```
program example
```

```
implicit none
complex*16::rho(0:1,0:1)
real*8::le
rho(0,0)=dcmplx(0.4d0,0.0d0)
rho(0,1)=dcmplx(-0.2d0,-0.1d0)
rho(1,0)=dcmplx(-0.2d0,0.1d0)
rho(1,1)=dcmplx(0.6d0,0.0d0)
call LEDMC(rho,2,le)
write(111,*)le
end program
```

The file "fort.111" contains the required linear entropy.

```
0.37999999999999989
```

There is a term called participation ratio which is also related to purity of state and it is defined as

$$PR = \frac{1}{Tr(\rho^2)}. \tag{4.25}$$

PR is interpreted as the effective number of pure states entering to form the mixture. We have given an example of finding the participation ratio for $\mathbb{I}/2$.

```
program example
real*8::I(0:1,0:1),ratio
complex*16::trace
I=0.0d0
I(0,0)=0.5d0
I(1,1)=0.5d0
call TRPOWNSR(I,2,2.0d0,trace)
ratio=1.0d0/real(trace)
print*,ratio
end program
```

which prints the ratio equal to 2.

4.10   Relative entropy

The closeness and distinguishability between two quantum states ρ and σ is quantified in terms of relative entropy [11, 48] and it can be defined as

$$S(\rho \parallel \sigma) = Tr(\rho \log \rho) - Tr(\rho \log \sigma). \tag{4.26}$$

Note that there is no upper limit on the value of relative entropy and $S(\rho \parallel \sigma) \geq 0$, the equality holds when $\rho = \sigma$. This definition of relative entropy is with respect to base e, however it can be changed to any base by dividing it with the appropriate conversion factor. Relative entropy is used in the calculation of quantum mutual information which is an important quantity in information theory.

4.10.1 Between two real density matrices

---

subroutine REDMR(A,B,n,re)

Parameters

In/Out	Argument	Description
[in]	n	n (integer) is the dimension of matrices A and B which are the states ρ and σ, respectively
[in]	A	A is a real*8 array of dimension (0:n−1,0:n−1)
[in]	B	B is a real*8 array of dimension (0:n−1,0:n−1)
[out]	re	re (real*8) is the relative entropy between states ρ and σ

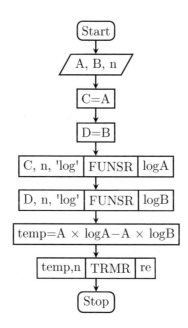

FIGURE 4.10: Flowchart of REDMR

Implementation

```
subroutine REDMR(A,B,n,re)
integer::n,i
real*8::A(0:n-1,0:n-1),B(0:n-1,0:n-1),re
real*8::logA(0:n-1,0:n-1),logB(0:n-1,0:n-1)
real*8::temp(0:n-1,0:n-1),C(0:n-1,0:n-1),D(0:n-1,0:n-1)
 C=A
D=B
call FUNSR(C,n,'log',logA)
```

```
call FUNSR(D,n,'log',logB)
temp=matmul(A,logA)-matmul(A,logB)
call TRMR(temp,n,re)
end subroutine
```

Example

In this example we are finding the relative entropy between the two-qubit generalized Werner states $\rho(0.3)$ and $\rho(0.5)$.

```
program example
implicit none
real*8::re
real*8::rho1(0:3,0:3),rho2(0:3,0:3)
integer::i,j
call NQBWERNER(2,0.3d0,rho1)
call NQBWERNER(2,0.5d0,rho2)
call REDMR(rho1,rho2,4,re)
write(111,*)re
end program
```

The file "fort.111" contains the required relative entropy.

```
4.6290422517800467E-002
```

4.10.2   Between two complex density matrices

```
subroutine REDMC(A,B,n,re)
```

Parameters

In/Out	Argument	Description
[in]	n	n (integer) is the dimension of matrices A and B which are the states ρ and σ, respectively
[in]	A	A is a complex*16 array of dimension $(0{:}n{-}1,0{:}n{-}1)$
[in]	B	B is a complex*16 array of dimension $(0{:}n{-}1,0{:}n{-}1)$
[out]	re	re (real*8) is the relative entropy between states ρ and σ

Implementation

```
subroutine REDMC(A,B,n,re)
integer::n,i
complex*16::A(0:n-1,0:n-1),B(0:n-1,0:n-1),re1
complex*16::logA(0:n-1,0:n-1),logB(0:n-1,0:n-1)
complex*16::temp(0:n-1,0:n-1),C(0:n-1,0:n-1),D(0:n-1,0:n-1)
real*8::re
```

```
C=A
D=B
call FUNHC(C,n,'log',logA)
call FUNHC(D,n,'log',logB)
temp=matmul(A,logA)-matmul(A,logB)
call TRMC(temp,n,re1)
re=real(re1)
end subroutine
```

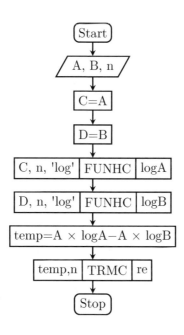

FIGURE 4.11: Flowchart of REDMC

Example

In this example we are finding the relative entropy between ρ_1 and ρ_2.

```
program example
implicit none
complex*16::rho1(0:1,0:1),rho2(0:1,0:1)
real*8::re
rho1(0,0)=dcmplx(0.4d0,0.0d0)
rho1(0,1)=dcmplx(-0.2d0,-0.1d0)
rho1(1,0)=dcmplx(-0.2d0,0.1d0)
rho1(1,1)=dcmplx(0.6d0,0.0d0)
rho2(0,0)=dcmplx(0.4d0,0.0d0)
rho2(0,1)=dcmplx(-0.1d0,-0.4d0)
rho2(1,0)=dcmplx(-0.1d0,0.4d0)
rho2(1,1)=dcmplx(0.6d0,0.0d0)
call REDMC(rho1,rho2,2,re)
write(111,*)re
end program
```

The file "fort.111" contains the required relative entropy.

```
0.34904660185936526
```

4.11   Trace distance

The trace distance [11, 49] is a metric on the space of density matrices and gives a measure of the distinguishability between two states. Trace distance between two density matrices ρ and σ is defined as follows,

$$T(\rho,\sigma) = \frac{1}{2} \parallel \rho - \sigma \parallel . \tag{4.27}$$

From the above, we see that the trace distance is directly related to the trace norm of the difference between the states.

4.11.1   Between two real density matrices

```
subroutine TRDISDMR(A,B,n,T)
```

Parameters

In/Out	Argument	Description
[in]	n	n (integer) is the dimension of matrices A and B which are the states ρ and σ, respectively
[in]	A	A is a real*8 array of dimension (0:n−1,0:n−1)
[in]	B	B is a real*8 array of dimension (0:n−1,0:n−1)
[out]	T	T (real*8) is the trace distance between states ρ and σ

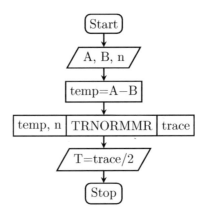

FIGURE 4.12: Flowchart of TRDISDMR

Implementation

```fortran
subroutine TRDISDMR(A,B,n,T)
implicit none
integer::n
real*8::A(0:n-1,0:n-1),B(0:n-1,0:n-1),T
real*8::temp(0:n-1,0:n-1),trace
temp=A-B
call TRNORMMR(temp,n,trace)
T=trace/2.0d0
end subroutine
```

Example

In this example we are finding the trace distance between the two-qubit generalized Werner states $\rho(0.3)$ and $\rho(0.5)$.

```fortran
program example
implicit none
real*8::rho1(0:3,0:3),T,rho2(0:3,0:3)
call NQBWERNER(2,0.3d0,rho1)
call NQBWERNER(2,0.5d0,rho2)
call TRDISDMR(rho1,rho2,4,T)
write(111,*)T
end program
```

The file "fort.111" contains the required trace distance.

```
0.14999999999999997
```

4.11.2 Between two complex density matrices

```fortran
subroutine TRDISDMC(A,B,n,T)
```

Parameters

In/Out	Argument	Description
[in]	n	n (integer) is the dimension of matrices A and B which are the states ρ and σ, respectively
[in]	A	A is a complex*16 array of dimension (0:n−1,0:n−1)
[in]	B	B is a complex*16 array of dimension (0:n−1,0:n−1)
[out]	T	T (real*8) is the trace distance between states ρ and σ

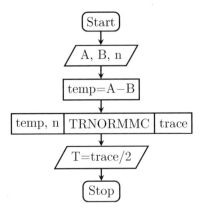

FIGURE 4.13: Flowchart of TRDISDMC

Implementation

```
subroutine TRDISDMC(A,B,n,T)
implicit none
integer::n,i
complex*16::A(0:n-1,0:n-1),B(0:n-1,0:n-1),T
complex*16::temp(0:n-1,0:n-1)
real*8::trace
temp=A-B
call TRNORMMC(temp,n,trace)
T=trace/2.0d0
end subroutine
```

Example

In this example we are finding the trace distance between ρ_1 and ρ_2.

```
program example
implicit none
complex*16::rho1(0:1,0:1),rho2(0:1,0:1)
real*8::T
rho1(0,0)=dcmplx(0.4d0,0.0d0)
rho1(0,1)=dcmplx(-0.2d0,-0.1d0)
rho1(1,0)=dcmplx(-0.2d0,0.1d0)
rho1(1,1)=dcmplx(0.6d0,0.0d0)
rho2(0,0)=dcmplx(0.4d0,0.0d0)
rho2(0,1)=dcmplx(-0.1d0,-0.4d0)
rho2(1,0)=dcmplx(-0.1d0,0.4d0)
rho2(1,1)=dcmplx(0.6d0,0.0d0)
call TRDISDMC(rho1,rho2,2,T)
write(111,*)T
end program
```

The file "fort.111" contains the required trace distance.

```
     0.31622776601683800
```

4.12 Fidelity

Fidelity [50, 51] is a measure of how close two quantum states are. Fidelity between two density matrices ρ and σ can be found using the trace norm as

$$F = \parallel \sqrt{\rho}\sqrt{\sigma} \parallel^2. \tag{4.28}$$

In the special case when ρ and σ are pure states, that is, $\rho = |\psi_\rho\rangle\langle\psi_\rho|$ and $\sigma = |\psi_\sigma\rangle\langle\psi_\sigma|$. Fidelity can be found as

$$F = |\langle\psi_\rho|\psi_\sigma\rangle|^2. \tag{4.29}$$

Fidelity between a pure state $|\psi\rangle$ and a density matrix σ can also be found as

$$F = \langle\psi|\sigma|\psi\rangle. \tag{4.30}$$

Note that fidelity always lies between 0 and 1. In fact the fidelity for pure states is similar to the overlap between the states.

4.12.1 Between two real states

```
subroutine FIDVR(psi1,psi2,n,F)
```

Parameters

In/Out	Argument	Description
[in]	n	n (integer) is the dimension of the vectors psi1 and psi2
[in]	psi1	psi1 is a real*8 array of dimension (0:n−1)
[in]	psi2	psi2 is a real*8 array of dimension (0:n−1)
[out]	F	F (real*8) is the fidelity between states psi1 and psi2

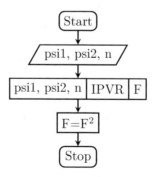

FIGURE 4.14: Flowchart of FIDVR

Implementation

```
subroutine FIDVR(psi1,psi2,n,F)
implicit none
integer::n,i
real*8::psi1(0:n-1),psi2(0:n-1),F
call IPVR(psi1,psi2,n,F)
F=F*F
end subroutine
```

Example

In this example we are finding the fidelity between two-qubit states $|\psi_1\rangle$ and $|\psi_2\rangle$.

```
program example
implicit none
integer::i
real*8::psi1(0:3),F,psir(0:3)
do i=0,3
    psir(i)=i+1
end do
call NORMALIZEVR(psir,4)
psi1=0.0d0
psi1(0)=1.0d0/sqrt(2.0d0)
psi1(3)=1.0d0/sqrt(2.0d0)
call FIDVR(psi1,psir,4,F)
write(111,*)F
end program
```

The file "fort.111" contains the required fidelity.

```
  0.41666666666666652
```

4.12.2   Between two complex states

```
subroutine FIDVC(psi1,psi2,n,F)
```

Parameters

In/Out	Argument	Description
[in]	n	n (integer) is the dimension of the vectors psi1 and psi2
[in]	psi1	psi1 is a complex*16 array of dimension (0:n−1)
[in]	psi2	psi2 is a complex*16 array of dimension (0:n−1)
[out]	F	F (real*8) is the fidelity between states psi1 and psi2

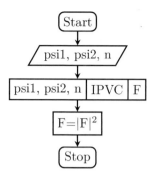

FIGURE 4.15: Flowchart of FIDVC

Implementation

```
subroutine FIDVC(psi1,psi2,n,F)
implicit none
integer::n,i
real*8::F
complex*16::psi1(0:n-1),psi2(0:n-1),F1
call IPVC(psi1,psi2,n,F1)
F1=F1*conjg(F1)
F=real(F1)
end subroutine
```

Example

In this example we are finding the fidelity between two-qubit states $|\psi_3\rangle$ and $|\psi_4\rangle$.

```
program example
implicit none
complex*16::psic1(0:3),psic2(0:3)
real*8::F
integer::i
do i=0,3
    psic1(i)=dcmplx(i,i+1)
end do
call NORMALIZEVC(psic1,4)
do i=0,3
    psic2(i)=dcmplx(i,i+2)
end do
call NORMALIZEVC(psic2,4)
call FIDVC(psic1,psic2,4,F)
write(111,*)F
end program
```

The file "fort.111" contains the required fidelity.

```
0.98663101604278070
```

4.12.3   Between two real density matrices

---
subroutine FIDDMR(rho1,rho2,n,F)

Parameters

In/Out	Argument	Description
[in]	n	n (integer) is the dimension of the matrices rho1 and rho2
[in]	rho1	rho1 is a real*8 array of dimension $(0{:}n{-}1,0{:}n{-}1)$
[in]	rho2	rho2 is a real*8 array of dimension $(0{:}n{-}1,0{:}n{-}1)$
[out]	F	F (real*8) is the fidelity between density matrices rho1 and rho2

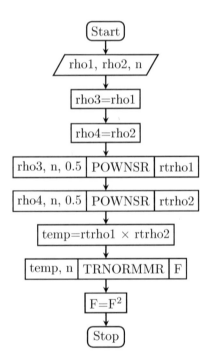

FIGURE 4.16: Flowchart of FIDDMR

Implementation

```
subroutine FIDDMR(rho1,rho2,n,F)
implicit none
integer::n
real*8::rho1(0:n-1,0:n-1),rho2(0:n-1,0:n-1),temp(0:n-1,0:n-1)
complex*16::rtrho1(0:n-1,0:n-1),rtrho2(0:n-1,0:n-1)
complex*16::temp1(0:n-1,0:n-1)
```

```
real*8::F,rho3(0:n-1,0:n-1),rho4(0:n-1,0:n-1)
real*8::pow
pow=0.5d0
rho3=rho1
rho4=rho2
call POWNSR(rho3,n,pow,rtrho1)
call POWNSR(rho4,n,pow,rtrho2)
temp1=matmul(rtrho1,rtrho2)
temp=real(temp1)
call TRNORMMR(temp,n,F)
F=F**2
end subroutine
```

Example

In this example we are finding the fidelity between two-qubit generalized Werner states $\rho(0.3)$ and $\rho(0.5)$.

```
program example
implicit none
integer::i,j
real*8::rho1(0:3,0:3),F,rho2(0:3,0:3)
call NQBWERNER(2,0.3d0,rho1)
call NQBWERNER(2,0.5d0,rho2)
call FIDDMR(rho1,rho2,4,F)
write(111,*)F
end program
```

The file "fort.111" contains the required fidelity.

```
0.97726738593353579
```

4.12.4 Between two complex density matrices

```
subroutine FIDDMC(rho1,rho2,n,F)
```

Parameters

In/Out	Argument	Description
[in]	n	n (integer) is the dimension of the matrices rho1 and rho2
[in]	rho1	rho1 is a complex*16 array of dimension (0:n−1,0:n−1)
[in]	rho2	rho2 is a complex*16 array of dimension (0:n−1,0:n−1)
[out]	F	F (real*8) is the fidelity between density matrices rho1 and rho2

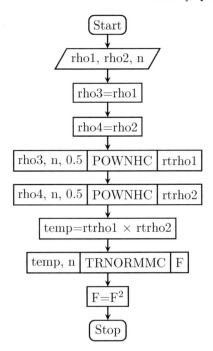

FIGURE 4.17: Flowchart of FIDDMC

Implementation

```
subroutine FIDDMC(rho1,rho2,n,F)
implicit none
complex*16::rho1(0:n-1,0:n-1),rho2(0:n-1,0:n-1)
complex*16::rtrho1(0:n-1,0:n-1),rtrho2(0:n-1,0:n-1),temp(0:n-1,0:n-1)
complex*16::rho3(0:n-1,0:n-1),rho4(0:n-1,0:n-1)
integer::n
real*8::pow,F
rho3=rho1
rho4=rho2
call POWNHC(rho3,n,0.5d0,rtrho1)
call POWNHC(rho4,n,0.5d0,rtrho2)
temp=matmul(rtrho1,rtrho2)
call TRNORMMC(temp,n,F)
F=F**2
end subroutine
```

Example

In this example we are finding the fidelity between the ρ_1 and ρ_2.

```
program example
implicit none
complex*16::rho1(0:1,0:1),rho2(0:1,0:1)
real*8::F
rho1(0,0)=dcmplx(0.4d0,0.0d0)
```

```
rho1(0,1)=dcmplx(-0.2d0,-0.1d0)
rho1(1,0)=dcmplx(-0.2d0,0.1d0)
rho1(1,1)=dcmplx(0.6d0,0.0d0)
rho2(0,0)=dcmplx(0.4d0,0.0d0)
rho2(0,1)=dcmplx(-0.1d0,-0.4d0)
rho2(1,0)=dcmplx(-0.1d0,0.4d0)
rho2(1,1)=dcmplx(0.6d0,0.0d0)
call FIDDMC(rho1,rho2,2,F)
write(111,*)F
end program
```

The file "fort.111" contains the required fidelity.

```
0.87065125189341619
```

4.12.5   Between a real state and a real density matrix

```
subroutine FIDVRDMR(n,rho,psi,F)
```

Parameters

In/Out	Argument	Description
[in]	n	n (integer) is the dimension of the matrices rho and psi
[in]	rho	rho is a real*8 array of dimension (0:n−1,0:n−1)
[in]	psi	psi is a real*8 array of dimension (0:n−1)
[out]	F	F (real*8) is the fidelity between psi and rho

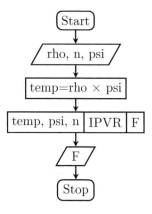

FIGURE 4.18: Flowchart of FIDVRDMR

Implementation

```
subroutine FIDVRDMR(n,rho,psi,F)
implicit none
integer::n,i
real*8::rho(0:n-1,0:n-1),psi(0:n-1),F,temp(0:n-1)
temp=matmul(rho,psi)
call IPVR(temp,psi,n,F)
end subroutine
```

Example

In this example we are finding the fidelity between two-qubit state $|\psi_2\rangle$ and two-qubit generalized Werner state $\rho(0.3)$.

```
program example
implicit none
integer::i,j
real*8::psi(0:3),rho(0:3,0:3),F
do i=0,3
    psi(i)=i+1
end do
call NORMALIZEVR(psi,4)
call NQBWERNER(2,0.3d0,rho)
call FIDVRDMR(4,rho,psi,F)
write(111,*)F
end program
```

The file "fort.111" contains the required fidelity.

```
0.29999999999999993
```

4.12.6   Between a complex state and a complex density matrix

```
subroutine FIDVCDMC(n,rho,psi,F)
```

Parameters

In/Out	Argument	Description
[in]	n	n (integer) is the dimension of the matrices rho and psi
[in]	rho	rho is a complex*16 array of dimension $(0:n-1,0:n-1)$
[in]	psi	psi is a complex*16 array of dimension $(0:n-1)$
[out]	F	F (real*8) is the fidelity between psi and rho

Implementation

```
subroutine FIDVCDMC(n,rho,psi,F)
implicit none
integer::n,i
complex*16::rho(0:n-1,0:n-1),psi(0:n-1),F1,temp(0:n-1)
real*8::F
temp=matmul(rho,psi)
call IPVC(temp,psi,n,F1)
F=real(F1)
end subroutine
```

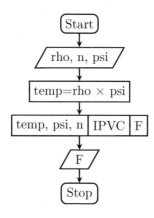

FIGURE 4.19: Flowchart of FIDVCDMC

Example

In this example we are finding the fidelity between the one-qubit state $|\psi_3\rangle$ and ρ_1.

```
program example
implicit none
complex*16::psi(0:1),rho(0:1,0:1)
real*8::F
integer::i,j
do i=0,1
    psi(i)=dcmplx(i,i+1)
end do
call NORMALIZEVC(psi,2)
rho(0,0)=dcmplx(0.4d0,0.0d0)
rho(0,1)=dcmplx(-0.2d0,-0.1d0)
rho(1,0)=dcmplx(-0.2d0,0.1d0)
rho(1,1)=dcmplx(0.6d0,0.0d0)
call FIDVCDMC(2,rho,psi,F)
write(111,*)F
end program
```

The file "fort.111" contains the required fidelity.

```
0.40000000000000008
```

4.13   Super fidelity

Superfidelity [52] is an upper bound on the fidelity of two quantum states which is defined by

$$G(\rho, \sigma) = Tr(\rho\sigma) + \sqrt{(1 - Tr(\rho^2)(1 - Tr(\sigma^2)},   \qquad (4.31)$$

for two-qubit states the super fidelity is the same as fidelity. In its own right, super fidelity can be used as an entanglement measure.

4.13.1   Between two real density matrices

subroutine SUPFIDDMR(rho1,rho2,n,F)

Parameters

In/Out	Argument	Description
[in]	n	n (integer) is the dimension of the matrices rho1 and rho2
[in]	rho1	rho1 is a real*8 array of dimension (0:n−1,0:n−1)
[in]	rho2	rho2 is a real*8 array of dimension (0:n−1,0:n−1)
[out]	F	F (real*8) is the super fidelity between density matrices rho1 and rho2

Implementation

```
subroutine SUPFIDDMR(rho1,rho2,n,F)
implicit none
integer::n
real*8::rho1(0:n-1,0:n-1),rho2(0:n-1,0:n-1)
real*8::a1(0:n-1,0:n-1),a2(0:n-1,0:n-1),a3(0:n-1,0:n-1)
real*8::F,trace1,trace2,trace3,temp1,temp2
a1=matmul(rho1,rho2)
a2=matmul(rho1,rho1)
a3=matmul(rho2,rho2)
call TRMR(a1,n,trace1)
call TRMR(a2,n,trace2)
call TRMR(a3,n,trace3)
temp1=1-trace2
if (abs(temp1) .le. 0.0000000000001) then
    temp1=0.0d0
end if
temp2=1-trace3
if (abs(temp2) .le. 0.0000000000001) then
    temp2=0.0d0
end if
F=trace1+dsqrt(temp1)*dsqrt(temp2)
end subroutine
```

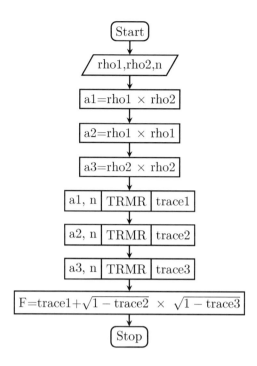

FIGURE 4.20: Flowchart of SUPFIDDMR

Example

In this example we are finding the super fidelity between the two-qubit generalized Werner states $\rho(0.3)$ and $\rho(0.5)$.

```
program example
implicit none
real*8::F
real*8::rho1(0:3,0:3),rho2(0:3,0:3)
call NQBWERNER(2,0.3d0,rho1)
call NQBWERNER(2,0.5d0,rho2)
call SUPFIDDMR(rho1,rho2,4,F)
write(111,*)F
end program
```

The file "fort.111" contains the required super fidelity.

```
0.98210168656968655
```

4.13.2 Between two complex density matrices

```
subroutine SUPFIDDMC(rho1,rho2,n,F)
```

Parameters

In/Out	Argument	Description
[in]	n	n (integer) is the dimension of the matrices rho1 and rho2
[in]	rho1	rho1 is a complex*16 array of dimension (0:n−1,0:n−1)
[in]	rho2	rho2 is a complex*16 array of dimension (0:n−1,0:n−1)
[out]	F	F (real*8) is the super fidelity between density matrices rho1 and rho2

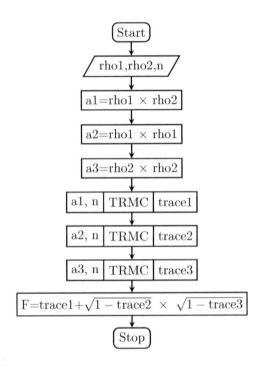

FIGURE 4.21: Flowchart of SUPFIDDMC

Implementation

```
subroutine SUPFIDDMC(rho1,rho2,n,F)
implicit none
integer::n
complex*16::rho1(0:n-1,0:n-1),rho2(0:n-1,0:n-1)
complex*16::a1(0:n-1,0:n-1),a2(0:n-1,0:n-1),a3(0:n-1,0:n-1)
complex*16::trace1,trace2,trace3
real*8::F,temp1,temp2
a1=matmul(rho1,rho2)
a2=matmul(rho1,rho1)
a3=matmul(rho2,rho2)
call TRMC(a1,n,trace1)
```

```
call TRMC(a2,n,trace2)
call TRMC(a3,n,trace3)
temp1=1-real(trace2)
if (abs(temp1) .le. 0.0000000000001) then
    temp1=0.0d0
end if
temp2=1-real(trace3)
if (abs(temp2) .le. 0.0000000000001) then
    temp2=0.0d0
end if
F=real(trace1)+dsqrt(temp1)*dsqrt(temp2)
end subroutine
```

Example

In this example we are finding the super fidelity between ρ_1 and ρ_2.

```
program example
implicit none
complex*16::rho1(0:1,0:1),rho2(0:1,0:1)
real*8::F
rho1(0,0)=dcmplx(0.4d0,0.0d0)
rho1(0,1)=dcmplx(-0.2d0,-0.1d0)
rho1(1,0)=dcmplx(-0.2d0,0.1d0)
rho1(1,1)=dcmplx(0.6d0,0.0d0)
rho2(0,0)=dcmplx(0.4d0,0.0d0)
rho2(0,1)=dcmplx(-0.1d0,-0.4d0)
rho2(1,0)=dcmplx(-0.1d0,0.4d0)
rho2(1,1)=dcmplx(0.6d0,0.0d0)
call SUPFIDDMC(rho1,rho2,2,F)
write(111,*)F
end program
```

The file "fort.111" contains the required super fidelity.

```
 0.87065125189341586
```

4.14   Bures distance

Bures distance [53–55] is defined as an infinitesimal distance between density matrix operators defining quantum states. Bures distance between two quantum states ρ and σ can be defined as,

$$D = \sqrt{2(1 - \sqrt{F(\rho,\sigma)})}, \tag{4.32}$$

where F is the fidelity between two quantum states ρ and σ. In its own right bures distance qualifies to be an entanglement measure.

4.14.1 Between two real states

---

subroutine BURDISVR(psi1,psi2,n,D)

Parameters

In/Out	Argument	Description
[in]	n	n (integer) is the dimension of the vectors psi1 and psi2
[in]	psi1	psi1 is a real*8 array of dimension (0:n−1)
[in]	psi2	psi2 is a real*8 array of dimension (0:n−1)
[out]	D	D (real*8) is the Bures distance between states psi1 and psi2

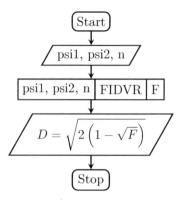

FIGURE 4.22: Flowchart of BURDISVR

Implementation

```
subroutine BURDISVR(psi1,psi2,n,D)
implicit none
integer::n
real*8::psi1(0:n-1),psi2(0:n-1),D
real*8::F,temp
call FIDVR(psi1,psi2,n,F)
temp=1-dsqrt(F)
If (temp .le. 0.0000000000001) then
temp=0.0d0
end if
D=dsqrt(2*(temp))
end subroutine
```

Example

In this example we are finding the Bures distance between the two-qubit states $|\psi_1\rangle$ and $|\psi_2\rangle$.

```
program example
implicit none
integer::i
real*8::psi2(0:3),D,psi1(0:3)
do i=0,3
    psi1(i)=i+1
end do
call NORMALIZEVR(psi1,4)
psi2=0.0d0
psi2(0)=1.0d0/sqrt(2.0d0)
psi2(3)=1.0d0/sqrt(2.0d0)
call BURDISVR(psi2,psi1,4,D)
write(111,*)D
end program
```

The file "fort.111" contains the required Bures distance.

```
0.84202467378586632
```

4.14.2 Between two complex states

```
subroutine BURDISVC(psi1,psi2,n,D)
```

Parameters

In/Out	Argument	Description
[in]	n	n (integer) is the dimension of the vectors psi1 and psi2
[in]	psi1	psi1 is a complex*16 array of dimension (0:n−1)
[in]	psi2	psi2 is a complex*16 array of dimension (0:n−1)
[out]	D	D (real*8) is the Bures distance between states psi1 and psi2

Implementation

```
subroutine BURDISVC(psi1,psi2,n,D)
implicit none
integer::n
complex*16::psi1(0:n-1),psi2(0:n-1)
real*8::D,F,temp
call FIDVC(psi1,psi2,n,F)
temp=1-dsqrt(F)
```

```
If (temp .le. 0.0000000000001) then
temp=0.0d0
end if
D=dsqrt(2*(temp))
end subroutine
```

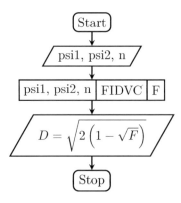

FIGURE 4.23: Flowchart of BURDISVC

Example

In this example we are finding the Bures distance between the two-qubit states $|\psi_3\rangle$ and $|\psi_4\rangle$.

```
program example
implicit none
complex*16::psic1(0:3),psic2(0:3)
real*8::D
integer::i
do i=0,3
    psic1(i)=dcmplx(i,i+1)
end do
call NORMALIZEVC(psic1,4)
do i=0,3
    psic2(i)=dcmplx(i,i+2)
end do
call NORMALIZEVC(psic2,4)
call BURDISVC(psic1,psic2,4,D)
write(111,*)D
end program
```

The file "fort.111" contains the required Bures distance.

```
0.11581868410942096
```

4.14.3   Between two real density matrices

```
subroutine BURDISDMR(rho1,rho2,n,D)
```

Parameters

In/Out	Argument	Description
[in]	n	n (integer) is the dimension of the matrices rho1 and rho2
[in]	rho1	rho1 is a real*8 array of dimension (0:n−1,0:n−1)
[in]	rho2	rho2 is a real*8 array of dimension (0:n−1,0:n−1)
[out]	D	D (real*8) is the Bures distance between density matrices rho1 and rho2

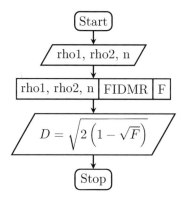

FIGURE 4.24: Flowchart of BURDISDMR

Implementation

```
subroutine BURDISDMR(rho1,rho2,n,D)
implicit none
integer::n
real*8::rho1(0:n-1,0:n-1),rho2(0:n-1,0:n-1),D
real*8::F,temp
call FIDDMR(rho1,rho2,n,F)
temp=1-dsqrt(F)
If (temp .le. 0.0000000000001) then
temp=0.0d0
end if
D=dsqrt(2*(temp))
end subroutine
```

Example

In this example we are finding the Bures distance between the two-qubit generalized Werner states $\rho(0.3)$ and $\rho(0.5)$.

```
program example
implicit none
```

```
integer::i,j
real*8::rho1(0:3,0:3),D,rho2(0:3,0:3)
call NQBWERNER(2,0.3d0,rho1)
call NQBWERNER(2,0.5d0,rho2)
call BURDISDMR(rho1,rho2,4,D)
write(111,*)D
end program
```

The file "fort.111" contains the required Bures distance.

```
0.15120613959059134
```

4.14.4 Between two complex density matrices

```
subroutine BURDISDMC(rho1,rho2,n,D)
```

Parameters

In/Out	Argument	Description
[in]	n	n (integer) is the dimension of the matrices rho1 and rho2
[in]	rho1	rho1 is a complex*16 array of dimension $(0{:}n{-}1,0{:}n{-}1)$
[in]	rho2	rho2 is a complex*16 array of dimension $(0{:}n{-}1,0{:}n{-}1)$
[out]	D	D (real*8) is the Bures distance between density matrices rho1 and rho2

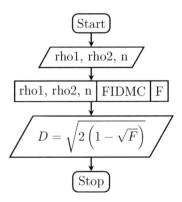

FIGURE 4.25: Flowchart of BURDISDMC

Implementation

```
subroutine BURDISDMC(rho1,rho2,n,D)
implicit none
```

```fortran
integer::n
complex*16::rho1(0:n-1,0:n-1),rho2(0:n-1,0:n-1)
real*8::D,F,temp
call FIDDMC(rho1,rho2,n,F)
temp=1-dsqrt(F)
If (temp .le. 0.0000000000001) then
temp=0.0d0
end if
D=dsqrt(2*(temp))
end subroutine
```

Example

In this example we are finding the Bures distance between ρ_1 and ρ_2.

```fortran
program example
implicit none
complex*16::rho1(0:1,0:1),rho2(0:1,0:1)
real*8::D
rho1(0,0)=dcmplx(0.4d0,0.0d0)
rho1(0,1)=dcmplx(-0.2d0,-0.1d0)
rho1(1,0)=dcmplx(-0.2d0,0.1d0)
rho1(1,1)=dcmplx(0.6d0,0.0d0)
rho2(0,0)=dcmplx(0.4d0,0.0d0)
rho2(0,1)=dcmplx(-0.1d0,-0.4d0)
rho2(1,0)=dcmplx(-0.1d0,0.4d0)
rho2(1,1)=dcmplx(0.6d0,0.0d0)
call BURDISDMC(rho1,rho2,2,D)
write(111,*)D
end program
```

The file "fort.111" contains the required Bures distance.

```
0.36582250433788543
```

4.15   Expectation value of an observable

In quantum mechanics, the expectation value [11, 56, 57] of any observable A with respect to pure state $|\psi\rangle$ is defined as

$$\langle A \rangle = \langle \psi | A | \psi \rangle. \tag{4.33}$$

The above equation also can be written if the state is given by a density matrix ρ as follows,

$$\langle A \rangle = Tr(\rho A). \tag{4.34}$$

When many measurements are done on N copies of the quantum system, the expectation value is the average value obtained when the number of such systems goes to infinity, that is $N \to \infty$. The expectation value of any observable is a real number.

4.15.1 For a real observable via a real state

---

subroutine AVGVRMR(n,A,psi,val)

Parameters

In/Out	Argument	Description
[in]	n	n (integer) is the dimension of the matrix A which is the observable
[in]	A	A is a real*8 array of dimension (0:n−1,0:n−1)
[in]	psi	psi is a real*8 array of dimension (0:n−1)
[out]	val	val (real*8) is the expectation value of A corresponding to the state psi

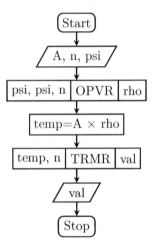

FIGURE 4.26: Flowchart of AVGVRMR

Implementation

```
subroutine AVGVRMR(n,A,psi,val)
implicit none
integer::n,i,j
real*8::A(0:n-1,0:n-1),rho(0:n-1,0:n-1),val,temp(0:n-1,0:n-1),psi(0:n-1)
call OPVR(psi,psi,n,n,rho)
temp=matmul(A,rho)
call TRMR(temp,n,val)
end subroutine
```

Example

In this example we are finding the expectation value of σ^z with respect to $|1\rangle$.

```
program example
implicit none
real*8::sigmaz(0:1,0:1),psi(0:1),val
call PAULIZ(sigmaz)
psi(0)=0.0
psi(1)=1.0
call AVGVRMR(2,sigmaz,psi,val)
write(111,*)val
end program
```

The file "fort.111" contains the required expectation value.

```
-1.0000000000000000
```

4.15.2 For a complex observable via a complex state

```
subroutine AVGVCMC(n,A,psi,val)
```

Parameters

In/Out	Argument	Description
[in]	n	n (integer) is the dimension of the matrix A which is the observable
[in]	A	A is a complex*16 array of dimension $(0{:}n{-}1,0{:}n{-}1)$
[in]	psi	psi is a complex*16 array of dimension $(0{:}n{-}1)$
[out]	val	val (real*8) is the expectation value of A corresponding to the state psi

Implementation

```
subroutine AVGVCMC(n,A,psi,val)
implicit none
integer::n,i,j
complex*16::A(0:n-1,0:n-1),rho(0:n-1,0:n-1)
complex*16::val1,temp(0:n-1,0:n-1),psi(0:n-1)
real*8::val
call OPVC(psi,psi,n,n,rho)
temp=matmul(A,rho)
call TRMC(temp,n,val1)
val=real(val1)
end subroutine
```

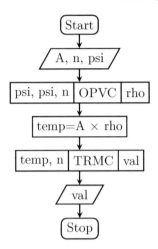

FIGURE 4.27: Flowchart of AVGVCMC

Example

In this example we are finding the expectation value of σ^y with respect to $\frac{|0\rangle - i|1\rangle}{2}$.

```
program example
implicit none
complex*16::sigmay(0:1,0:1),i,psi(0:1)
real*8::val
i=dcmplx(0,1)
call PAULIY(sigmay)
psi(0)=1/sqrt(2.0d0)
psi(1)=-i/sqrt(2.0d0)
call AVGVCMC(2,sigmay,psi,val)
write(111,*)val
end program
```

The file "fort.111" contains the required expectation value.

```
-0.99999999999999978
```

4.15.3   For a real observable via a real density matrix

```
subroutine AVGDMRMR(n,A,rho,val)
```

Parameters

In/Out	Argument	Description
[in]	n	n (integer) is the dimension of the matrix A which is the observable
[in]	A	A is a real*8 array of dimension (0:n−1,0:n−1)

In/Out	Argument	Description
[in]	rho	rho is a real*8 array of dimension (0:n−1,0:n−1)
[out]	val	val (real*8) is the expectation value of A corresponding to the density matrix rho

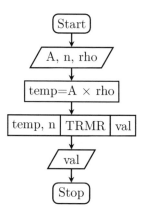

FIGURE 4.28: Flowchart of AVGDMRMR

Implementation

```
subroutine AVGDMRMR(n,A,rho,val)
implicit none
integer::n,i
real*8::A(0:n-1,0:n-1),rho(0:n-1,0:n-1),val,temp(0:n-1,0:n-1)
temp=matmul(rho,A)
call TRMR(temp,n,val)
end subroutine
```

Example

In this example we are finding the expectation value of σ^z with respect to $|0\rangle\langle0|$.

```
program example
implicit none
real*8::sigmaz(0:1,0:1),rho(0:1,0:1),val
call PAULIZ(sigmaz)
rho(0,:)=(/1,0/)
rho(1,:)=(/0,0/)
call AVGDMRMR(2,sigmaz,rho,val)
write(111,*)val
end program
```

The file "fort.111" contains the required expectation value.

```
1.0000000000000000
```

4.15.4 For a complex observable via a complex density matrix

---

subroutine AVGDMCMC(n,A,rho,val)

Parameters

In/Out	Argument	Description
[in]	n	n (integer) is the dimension of the matrix A which is the observable
[in]	A	A is a complex*16 array of dimension (0:n−1,0:n−1)
[in]	rho	rho is a complex*16 array of dimension (0:n−1,0:n−1)
[out]	val	val (real*8) is the expectation value of A corresponding to the density matrix rho

Implementation

```
subroutine AVGDMCMC(n,A,rho,val)
implicit none
integer::n,i
complex*16::A(0:n-1,0:n-1),rho(0:n-1,0:n-1),temp(0:n-1,0:n-1)
complex*16::val1
real*8::val
temp=matmul(rho,A)
call TRMC(temp,n,val1)
val=real(val1)
end subroutine
```

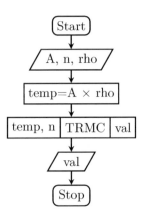

FIGURE 4.29: Flowchart of AVGDMCMC

Example

In this example we are finding the expectation value of σ^y with respect to $\left(\frac{|0\rangle + i|1\rangle}{\sqrt{2}}\right)\left(\frac{\langle 0| - i\langle 1|}{\sqrt{2}}\right)$.

```
program example
implicit none
complex*16::sigmay(0:1,0:1),rho(0:1,0:1),i,psi(0:1)
real*8::val
integer::x,y
i=dcmplx(0,1)
call PAULIY(sigmay)
psi(0)=1/sqrt(2.0d0)
psi(1)=i/sqrt(2.0d0)
do x=0,1
    do y=0,1
        rho(x,y)=psi(x)*conjg(psi(y))
    end do
end do
call AVGDMCMC(2,sigmay,rho,val)
write(111,*)val
end program
```

The file "fort.111" contains the required expectation value.

```
0.99999999999999978
```

4.16   The single-qubit measurement

For single qubit, there are two projective measurement operators [11, 58, 59] namely $M_1 = |0\rangle\langle 0|$ and $M_2 = |1\rangle\langle 1|$. The probability of getting an outcome i on measuring the observable M_i $\{i = 1, 2\}$ for the system in the pre measurement state $|\psi\rangle$ is given by,

$$p(i) = \langle\psi|M_i^\dagger M_i|\psi\rangle. \tag{4.35}$$

The normalized state after the measurement is given by,

$$|\psi'\rangle = \frac{M_i|\psi\rangle}{\sqrt{\langle\psi|M_i^\dagger M_i|\psi\rangle}}. \tag{4.36}$$

If the state of the system is defined by the density matrix ρ before the measurement, we get the probability of getting an outcome i on measuring the observable M_i as,

$$p(i) = Tr(M_i^\dagger M_i\rho), \tag{4.37}$$

and the normalized state after the measurement is given by,

$$\rho' = \frac{M_i\rho M_i^\dagger}{Tr(M_i^\dagger M_i\rho)}. \tag{4.38}$$

Note that if we have an N-qubit state, and we like to measure an observable with respect to some qubit, the recipe given can handle that too.

4.16.1 For a real state

```
subroutine MEASUREVR(s,x,psi,o,psim,pm)
```

Parameters

In/Out	Argument	Description
[in]	s	s (integer) is the total number of qubits
[in]	x	x (integer) is the index of the qubit to be measured
[in]	psi	psi is a real*8 array of dimension (0:2**s−1) which is the pre-measurement state
[in]	o	o (character*1) is a string defining the type of measurement operator if o = '0', it measures the nth qubit in the up state if o = '1', it measures the nth qubit in the down state
[out]	psim	psim is a real*8 array of dimension (0:2**s−1), which is the normalized post measurement state
[out]	pm	pm (real*8) is the value of probability for getting the state psim

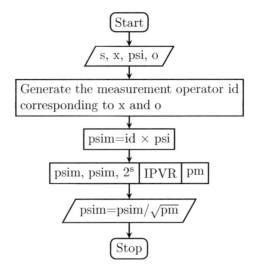

FIGURE 4.30: Flowchart of MEASUREVR

Implementation

```
subroutine MEASUREVR(s,x,psi,o,psim,pm)
implicit none
```

```
integer::s,n,i,j,x
character::o
real*8::psi(0:2**s-1),psim(0:2**s-1),pm,id(0:2**s-1,0:2**s-1)
id=0.0d0
n=x+1
if (o=='0') then
    do i=0,2**s-1,2**(s-n+1)
        do j=i,i+2**(s-n)-1
            id(j,j)=1.0d0
        end do
    end do
end if
if (o=='1') then
    do i=2**(s-n),2**s-1,2**(s-n+1)
        do j=i,i+2**(s-n)-1
            id(j,j)=1.0d0
        end do
    end do
end if
psim=matmul(id,psi)
call IPVR(psim,psim,2**s,pm)
psim=psim/sqrt(pm)
end subroutine
```

Example

In this example we are measuring the second qubit to be in the up state of the given state $\frac{|01\rangle - |10\rangle}{\sqrt{2}}$.

```
program example
implicit none
integer,parameter::s=2,n=1
integer::i
real*8::psi(0:2**s-1),psim(0:2**s-1),pm
psi=0.0d0
psi(1)=1/sqrt(2.0d0)
psi(2)=-1/sqrt(2.0d0)
call MEASUREVR(s,n,psi,'0',psim,pm)
do i=0,3
write(111,*)psim(i)
end do
write(111,*)pm
end program
```

The file "fort.111" contains the required post measurement state and the probability.

```
 0.0000000000000000
 0.0000000000000000
-1.0000000000000000
 0.0000000000000000
 0.49999999999999989
```

4.16.2   For a complex state

---

subroutine MEASUREVC(s,x,psi,o,psim,pm)

Parameters

In/Out	Argument	Description
[in]	s	s (integer) is the total number of qubits
[in]	x	x (integer) is the index of the qubit to be measured
[in]	psi	psi is a complex*16 array of dimension (0:2**s−1) which is the pre-measurement state
[in]	o	o (character*1) is a string defining the type of measurement operator if o = '0', it measures the nth qubit in the up state if o = '1', it measures the nth qubit in the down state
[out]	psim	psim is a complex*16 array of dimension (0:2**s−1), which is the normalized post measurement state
[out]	pm	pm (real*8) is the value of probability for getting the state psim

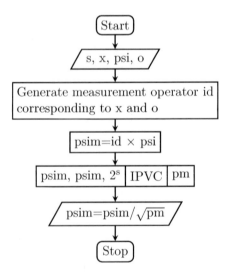

FIGURE 4.31: Flowchart of MEASUREVC

Implementation

```
subroutine MEASUREVC(s,x,psi,o,psim,pm)
implicit none
integer::s,n,i,j,x
character::o
complex*16::psi(0:2**s-1),psim(0:2**s-1),pm1
real*8::pm,id(0:2**s-1,0:2**s-1)
id=0.0d0
n=x+1
if (o=='0') then
    do i=0,2**s-1,2**(s-n+1)
        do j=i,i+2**(s-n)-1
            id(j,j)=1.0d0
        end do
    end do
end if
if (o=='1') then
    do i=2**(s-n),2**s-1,2**(s-n+1)
        do j=i,i+2**(s-n)-1
            id(j,j)=1.0d0
        end do
    end do
end if
psim=matmul(id,psi)
call IPVC(psim,psim,2**s,pm1)
pm=real(pm1)
psim=psim/sqrt(pm)
end subroutine
```

Example

In this example we are measuring the second qubit to be in the up state of the given state $\frac{i|01\rangle - i|10\rangle}{\sqrt{2}}$.

```
program example
implicit none
integer,parameter::s=2,n=1
complex*16::psi(0:2**s-1),psim(0:2**s-1)
integer::i
real*8::pm
psi=0.0d0
psi(1)=dcmplx(0,1/sqrt(2.0d0))
psi(2)=dcmplx(0,-1/sqrt(2.0d0))
call MEASUREVC(s,n,psi,'0',psim,pm)
do i=0,3
    write(111,*)psim(i)
end do
write(111,*)pm
end program
```

The file "fort.111" contains the required post measurement state and the probability.

$$(0.0000000000000000, 0.0000000000000000)$$
$$(0.0000000000000000, 0.0000000000000000)$$
$$(0.0000000000000000, -1.0000000000000000)$$
$$(0.0000000000000000, 0.0000000000000000)$$
$$0.49999999999999989$$

4.16.3 For a real density matrix

subroutine MEASUREDMR(s,x,rho,o,rhom,pm)

Parameters

In/Out	Argument	Description
[in]	s	s (integer) is the total number of qubits
[in]	x	x (integer) is the index of the qubit to be measured
[in]	rho	rho is a real*8 array of dimension $(0{:}2^{**}\text{s}{-}1, 0{:}2^{**}\text{s}{-}1)$ which is the pre-measurement state
[in]	o	o (character*1) is a string defining the type of measurement operator if o = '0', it measures the nth qubit in the up state if o = '1', it measures the nth qubit in the down state
[out]	rhom	rhom is a real*8 array of dimension $(0{:}2^{**}\text{s}{-}1, 0{:}2^{**}\text{s}{-}1)$, which is the post measurement state
[out]	pm	pm (real*8) is the value of probability for getting the state rhom

Implementation

```
subroutine MEASUREDMR(s,x,rho,o,rhom,pm)
implicit none
integer::s,n,i,j,x
character::o
real*8::rho(0:2**s-1,0:2**s-1),rhom(0:2**s-1,0:2**s-1),pm
real*8::id(0:2**s-1,0:2**s-1)
id=0.0d0
n=x+1
if (o=='0') then
    do i=0,2**s-1,2**(s-n+1)
        do j=i,i+2**(s-n)-1
            id(j,j)=1.0d0
        end do
    end do
```

```
end if
if (o=='1') then
    do i=2**(s-n),2**s-1,2**(s-n+1)
        do j=i,i+2**(s-n)-1
            id(j,j)=1.0d0
        end do
    end do
end if
rhom=matmul(id,matmul(rho,transpose(id)))
call TRMR(rhom,2**s,pm)
rhom=rhom/pm
end subroutine
```

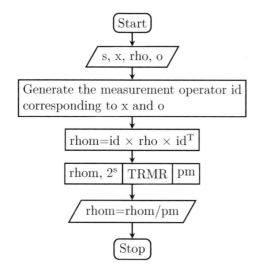

FIGURE 4.32: Flowchart of MEASUREDMR

Example

In this example we are measuring the second qubit to be in the up state of the given density matrix $\left(\frac{|01\rangle-|10\rangle}{\sqrt{2}}\right)\left(\frac{\langle 01|-\langle 10|}{\sqrt{2}}\right)$.

```
program example
implicit none
integer,parameter::s=2,n=1
real*8::psi(0:2**s-1),rho(0:2**s-1,0:2**s-1),pm
real*8::rhom(0:2**s-1,0:2**s-1)
integer::i,j
psi=0.0d0
psi(1)=1/sqrt(2.0d0)
psi(2)=-1/sqrt(2.0d0)
do i=0,2**s-1
    do j=0,2**s-1
        rho(i,j)=psi(i)*psi(j)
    end do
```

```
end do
call MEASUREDMR(s,n,rho,'0',rhom,pm)
do i=0,3
   write(111,*)int(rhom(i,:))
end do
write(111,*)pm
end program
```

The file "fort.111" contains the required post measurement state and the probability.

0	0	0	0
0	0	0	0
0	0	1	0
0	0	0	0

0.49999999999999989

4.16.4 For a complex density matrix

```
subroutine MEASUREDMC(s,x,rho,o,rhom,pm)
```

Parameters

In/Out	Argument	Description
[in]	s	s (integer) is the total number of qubits
[in]	x	x (integer) is the index of the qubit to be measured
[in]	rho	rho is a complex*16 array of dimension $(0:2**s-1,0:2**s-1)$ which is the pre-measurement state
[in]	o	o (character*1) is a string defining the type of measurement operator if o = '0', it measures the nth qubit in the up state if o = '1', it measures the nth qubit in the down state
[out]	rhom	rhom is a complex*16 array of dimension $(0:2**s-1,0:2**s-1)$, which is the post measurement state
[out]	pm	pm (real*8) is the value of probability for getting the state rhom

Implementation

```
subroutine MEASUREDMC(s,x,rho,o,rhom,pm)
implicit none
integer::s,n,i,j,x
character::o
complex*16::rho(0:2**s-1,0:2**s-1),rhom(0:2**s-1,0:2**s-1),pm1
real*8::id(0:2**s-1,0:2**s-1),pm
```

```
id=0.0d0
n=x+1
if (o=='0') then
    do i=0,2**s-1,2**(s-n+1)
        do j=i,i+2**(s-n)-1
            id(j,j)=1.0d0
        end do
    end do
end if
if (o=='1') then
    do i=2**(s-n),2**s-1,2**(s-n+1)
        do j=i,i+2**(s-n)-1
            id(j,j)=1.0d0
        end do
    end do
end if
rhom=matmul(id,matmul(rho,transpose(id)))
call TRMC(rhom,2**s,pm1)
rhom=rhom/pm1
pm=real(pm1)
end subroutine
```

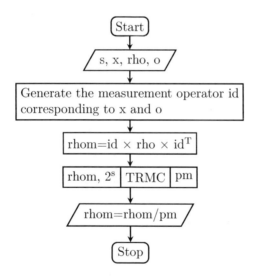

FIGURE 4.33: Flowchart of MEASUREDMC

Example

In this example we are measuring the second qubit to be in the up state of the given density matrix $\left(\frac{i|01\rangle - i|10\rangle}{\sqrt{2}}\right)\left(\frac{-i\langle01| + i\langle10|}{\sqrt{2}}\right)$.

```
program example
implicit none
integer,parameter::s=2,n=1
complex*16::psi(0:2**s-1),rho(0:2**s-1,0:2**s-1),rhom(0:2**s-1,0:2**s-1)
```

```
real*8::pm
integer::i,j
psi=0.0d0
psi(1)=dcmplx(0,1/sqrt(2.0d0))
psi(2)=dcmplx(0,-1/sqrt(2.0d0))
do i=0,2**s-1
    do j=0,2**s-1
        rho(i,j)=psi(i)*conjg(psi(j))
    end do
end do
call MEASUREDMC(s,n,rho,'0',rhom,pm)
do i=0,3
    write(111,*)int(rhom(i,:))
end do
write(111,*)pm
end program
```

The file "fort.111" contains the required post measurement state and the probability.

0	0	0	0
0	0	0	0
0	0	1	0
0	0	0	0

0.49999999999999989

5

Quantum Entanglement and Its Quantification

Quantum entanglement [60–64] is an exclusive property of quantum states, such a phenomenon is not there in classical mechanics. Entanglement means non-local correlations which exist between systems which are separated in space. Any measurement in one of the subsystems will affect the state of the other subsystem, essentially it means that the subsystems are no longer independent [65]. Entanglement plays an important role in not only understanding the basic nature of quantum states but also indispensable in understanding quantum states of condensed matter systems like spin chains, magnets, etc. [66]. Not only that, it is also used as a signature of quantum phase transitions [67] and also to perform major quantum information tasks which reduces the computational time compared to the classical methods. In this chapter, we shall see various important operations which are performed on quantum states like partial tracing and partial transpose. Not only that, we will also see how these operations can be used to obtain important entanglement measures [68] which is a quantitative description of entanglement. The nature of entanglement in pure and mixed states are very different from each other and as we go along we shall see the ways to detect and quantify entanglement.

As far as pure states are concerned, if we have a composite system AB which is in a state $|\psi\rangle_{AB}$ made of two subsystems A and B, if the state of the composite system can be written as a tensor product of the individual subsystem states then we say that $|\psi\rangle_{AB}$ unentangled, if not, they are entangled. For example consider one of the well known two-qubit Bell states which is,

$$
\begin{aligned}
|b_1\rangle \;&=\; \frac{1}{\sqrt{2}}(|00\rangle_{AB} + |11\rangle_{AB}) \\
&\neq\; |\text{some state of system } A\rangle \otimes |\text{some state of system } B\rangle
\end{aligned}
\tag{5.1}
$$

In the above equation, we see that $|b_1\rangle$ is an entangled state. However, in the same token, it is easy to see that the state $|\phi\rangle = \frac{1}{\sqrt{2}}(|00\rangle_{AB} + |01\rangle_{AB}) = \frac{1}{\sqrt{2}}(|0\rangle_A \otimes (|0\rangle_B + |1\rangle_B))$ is an unentangled state. In the state defined as $|b_1\rangle$, we see that it is defined in terms of the eigenvectors of the Pauli Z matrix, however, if we rotate the basis such that the state is written in terms of the eigen vectors of the Pauli X matrix, even then, the non-local correlations or entanglement does not die as the new state preserves these correlations upto an arbitrary global phase. In the language of measurement theory, we see that in the state $|b_1\rangle$, if we measure the spin in the first qubit to be in the state $|0\rangle$ we can immediately, without even measuring the second qubit say with 100 percent probability that it will be in the state $|0\rangle$, however, if we see the state $|\phi\rangle$, measuring the first qubit to be in state $|0\rangle$, will throw the state of the second qubit in a superposition namely $\frac{1}{\sqrt{2}}(|0\rangle_B + |1\rangle_B)$, which is a hallmark of entanglement.

However coming to the arena of density matrices, A density matrix ρ is called separable [69] if it can be written as a convex sum of product states ρ^A and ρ^B, as

$$
\rho = \sum_i p_i \rho_i^A \otimes \rho_i^B,
\tag{5.2}
$$

This definition coupled with the fact that there is no unique decomposition [70] for a given density matrix in terms of pure states leads to a huge market of entanglement measures and associated entanglement detection techniques. Some of them like PPT and reduction are commonly used and day by day the research on these increase the number of such measures and detection methods which practically makes it weaker or stronger with respect to each other [71]. For qubit-qubit and qubit-qutrit systems (2×2 and 2×3, respectively) the simple Peres–Horodecki criterion or the famous positive partial transpose PPT criteria [69] provides both a necessary and a sufficient criterion for separability. This cannot be generalized to higher-dimensional quantum states this directly making detection non-unique. In this chapter, we will be using the same states and density matrices as given in chapter 4 for demonstrating our recipes and outputs. In this chapter, the array named 'site' stores the index corresponding to the site of the qubits which has been done via zero-based indexing meaning 0 corresponds to first qubit and so on.

5.1   Partial trace

Partial trace [11, 72] is useful when we want to study the subsystems of a given composite system. Consider that, we have a bipartite composite system in the Hilbert space $H = H_A \otimes H_B$ and the density matrix corresponding to the composite system is ρ_{AB}. From this, if we are interested in knowing the subsystem states, either for subsystems A or B it can be done using the operation called partial tracing. The state of the subsystem is given by the reduced density matrix ρ_A. It can be found as follows,

$$\rho_A = Tr_B(\rho_{AB}). \tag{5.3}$$

Similarly, state of the subsystem B which is ρ_B is given by,

$$\rho_B = Tr_A(\rho_{AB}). \tag{5.4}$$

It can be generalized to multi qubit systems also in this way, if we have a N-qubit state $\rho_{123\cdots N}$, then if we are interested to know the state of the even qubit subsystem we can write $\rho_{246\cdots} = Tr_{135\cdots}(\rho_{123\cdots N})$. This implies, partial tracing can be done over a random set of qubits. The partial trace operation is the faithful reflection of the marginal of a multivariate probability distribution in the quantum domain. Partial trace operation is practically used in calculating most of the entanglement measures as we will see further down. In the recipes which we have given, for a N-qubit state, we can calculate the partial trace over any random set of qubits as $\rho_{i_1 i_2 \cdots i_N}$ provided $i_1 < i_2 < \cdots < i_N$. Note that flowchart of partial tracing for real and complex density matrices are the same.

5.1.1   Prelude to partial tracing – rearrangement of indices

The flowchart for doing the partial tracing is divided into two parts, in the first part we do a rearrangement of indices and in the second part, the construction of the reduced density matrix. The rearrangement procedure is common for all the routines involving the computation of partial trace. The output of first part will be stored in array called "ind". It stores the integer number from 0 to $2^N - 1$ in a rearranged manner, and this rearrangement depends upon the number of qubits to be traced out and the site of the qubits. This "ind" will be fed into the second part of the algorithm as shown in the dotted boxes.

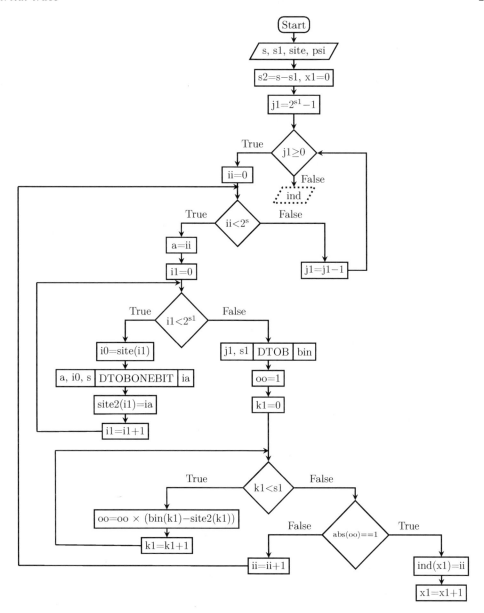

FIGURE 5.1: Flowchart of rearrangement of indices

5.1.2 For a real state

```
subroutine PTRVR(s,s1,site,vin,rdm)
```

Parameters

In/Out	Argument	Description
[in]	s	s (integer) is the total number of qubits

In/Out	Argument	Description
[in]	s1	s1 (integer) is the total number of qubits not to be traced out
[in]	site	site is an integer array of dimension $(0{:}s1-1)$ which contains the site indices of the qubits not to be traced out
[in]	vin	vin is a real*8 array of dimension $(0{:}2^{**}s-1)$ which is the input state
[out]	rdm	rdm is a real*8 array of dimension $(0{:}2^{**}s1-1,0{:}2^{**}s1-1)$ which is the required reduced density matrix

Implementation

```
subroutine PTRVR(s,s1,site,vin,rdm)
implicit none
integer::s, s1,s2
real*8::trace
real*8,dimension(0:2**s-1)::vin
real*8,dimension(0:2**s1-1,0:2**s1-1)::rdm
integer::ii,i1,a,i0,ia,j1,oo,k1,x1,y1,j,t1,t2,z1,i
integer,dimension(0:s1-1)::site,site2
integer,dimension(0:s1-1)::bin
integer,dimension(0:(2**(s-s1))*(2**s1)-1)::ind
s2=s-s1
x1=0
y1=0
ind=0
do j1=2**s1-1,0,-1
        do ii=0,2**s-1
                a=ii
                do i1=0,s1-1,1
                    i0=site(i1)
                    call DTOBONEBIT(a,ia,i0,s)
                    site2(i1)=ia
                enddo
                call DTOB(j1,bin,s1)
                oo=1
                do k1=0,s1-1,1
                oo=oo*(bin(k1)-site2(k1))
                end do
                if(abs(oo)==1) then
                    ind(x1)=ii
                    x1=x1+1
                end if
        end do
end do
```

```
rdm=0.0d0
do t1=0,2**s1-1,1
   do t2=0,2**s1-1,1
      do z1=0,2**s2-1,1
         rdm(t1,t2)=rdm(t1,t2)+vin(ind((2**s2)*t1+z1)) &
                         *vin(ind((2**s2)*t2+z1))
      enddo
   end do
end do
end subroutine
```

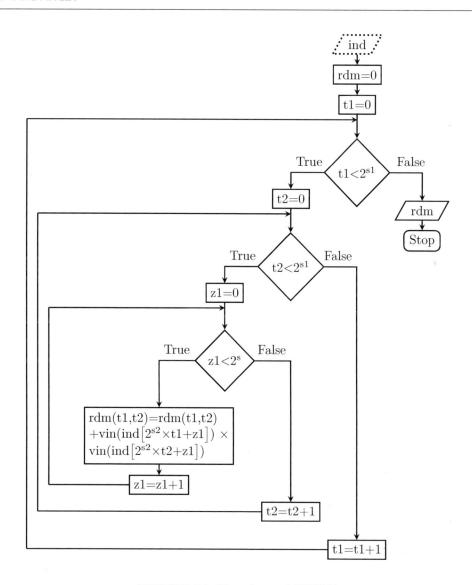

FIGURE 5.2: Flowchart of PTRVR

Example

In this example we are performing the partial trace on the four-qubit state $|\psi_2\rangle$ and we will get ρ_{24} as an output by tracing over qubits 1 and 3.

```
program example
implicit none
integer,parameter::s=4,s1=2
integer::i,j,site(0:s1-1)
real*8::psir(0:2**s-1),rdm(0:2**s1-1,0:2**s1-1)
do i=0,2**s-1
    psir(i)=i+1
end do
call NORMALIZEVR(psir,2**s)
site(0)=1
site(1)=3
call PTRVR(s,s1,site,psir,rdm)
do i=0,2**s1-1
    write(111,*)real(rdm(i,:))
end do
end program
```

The file "fort.111" contains the entries of the required reduced density matrix ρ_{24}.

0.141711235	0.157754004	0.205882356	0.221925139
0.157754004	0.176470593	0.232620314	0.251336902
0.205882356	0.232620314	0.312834233	0.339572191
0.221925139	0.251336902	0.339572191	0.368983954

5.1.3   For a complex state

```
subroutine PTRVC(s,s1,site,vin,rdm)
```

Parameters

In/Out	Argument	Description
[in]	s	s (integer) is the total number of qubits
[in]	s1	s1 (integer) is the total number of qubits not to be traced out
[in]	site	site is an integer array of dimension $(0{:}s1-1)$ which contains the site indices of the qubits not to be traced out
[in]	vin	vin is a complex*16 array of dimension $(0{:}2**s-1)$ which is the input state
[out]	rdm	rdm is a complex*16 array of dimension $(0{:}2**s1-1,0{:}2**s1-1)$ which is the required reduced density matrix

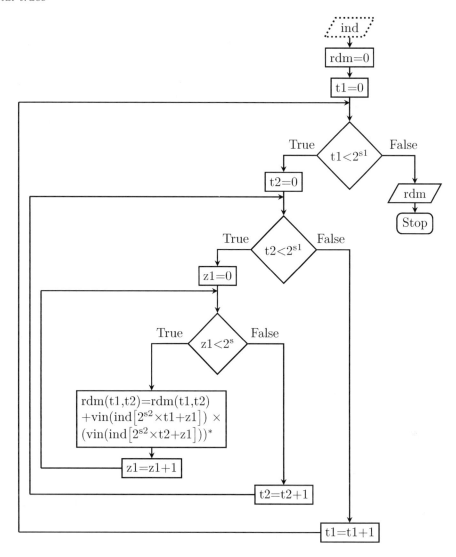

FIGURE 5.3: Flowchart of PTRVC

Implementation

```
subroutine PTRVC(s,s1,site,vin,rdm)
implicit none
integer::s, s1,s2
real*8::trace
complex*16,dimension(0:2**s-1)::vin
complex*16,dimension(0:2**s1-1,0:2**s1-1)::rdm
integer::ii,i1,a,i0,ia,j1,oo,k1,x1,y1,j,t1,t2,z1,i
integer,dimension(0:s1-1)::site,site2
integer,dimension(0:s1-1)::bin
integer,dimension(0:(2**(s-s1))*(2**s1)-1)::ind
s2=s-s1
x1=0
```

```
y1=0
ind=0

do j1=2**s1-1,0,-1
        do ii=0,2**s-1
                  a=ii
                  do i1=0,s1-1,1
                      i0=site(i1)
                      call DTOBONEBIT(a,ia,i0,s)
                      site2(i1)=ia
                  enddo
                  call DTOB(j1,bin,s1)
                  oo=1
                  do k1=0,s1-1,1
                  oo=oo*(bin(k1)-site2(k1))
                  end do
                  if(abs(oo)==1) then
                      ind(x1)=ii
                          x1=x1+1
                  end if
          end do
end do
rdm=0.0d0
do t1=0,2**s1-1,1
   do t2=0,2**s1-1,1
     do z1=0,2**s2-1,1
        rdm(t1,t2)=rdm(t1,t2)+vin(ind((2**s2)*t1+z1))* &
                       conjg(vin(ind((2**s2)*t2+z1)))
      enddo
   end do
end do
end subroutine
```

Example

In this example we are performing partial trace on the four-qubit state $|\psi_3\rangle$ and we will get the reduced density matrix ρ_4 as an output by tracing over qubits 1, 2 and 3.

```
program example
implicit none
integer,parameter::s=4,s1=1
integer::i,j,site(0:0)
complex*16::psic(0:2**s-1),rdm(0:2**s1-1,0:2**s1-1)
real*8::be
do i=0,2**s-1
    psic(i)=dcmplx(i,i+1)
end do
call NORMALIZEVC(psic,2**s)
site(0)=3
call PTRVC(s,s1,site,psic,rdm)
do i=0,2**s1-1
```

```
    do j=0,2**s1-1
        write(111,*)rdm(i,j)
    end do
end do
end program
```

The file "fort.111" contains the entries of required reduced density matrix ρ_4.

```
(0.45321637426900580,0.0000000000000000)
(0.49707602339181289,2.92397660818712437E-003)
(0.49707602339181289,-2.92397660818712437E-003)
(0.54678362573099415,0.0000000000000000)
```

5.1.4   For a real density matrix

```
subroutine PTRDMR(s,s1,site,rho,rdm)
```

Parameters

In/Out	Argument	Description
[in]	s	s (integer) is the total number of qubits
[in]	s1	s1 (integer) is the total number of qubits not to be traced out
[in]	site	site is an integer array of dimension $(0{:}s1-1)$ which contains the site indices of the qubits not to be traced out
[in]	rho	rho is a real*8 array of dimension $(0{:}2^{**}s{-}1,0{:}2^{**}s{-}1)$ which is the input density matrix
[out]	rdm	rdm is a real*8 array of dimension $(0{:}2^{**}s1{-}1,0{:}2^{**}s1{-}1)$ which is the required reduced density matrix

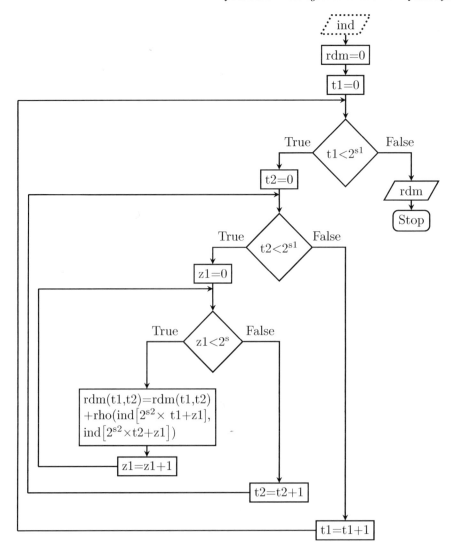

FIGURE 5.4: Flowchart of PTRDMR and PTRDMC

Implementation

```
subroutine PTRDMR(s,s1,site,rho,rdm)
implicit none
integer::s, s1,s2
real*8::trace
real*8,dimension(0:2**s-1,0:2**s-1)::rho
real*8,dimension(0:2**s1-1,0:2**s1-1)::rdm
integer::ii,i1,a,i0,ia,j1,oo,k1,x1,y1,j,t1,t2,z1,i
integer,dimension(0:s1-1)::site,site2
integer,dimension(0:s1-1)::bin
integer,dimension(0:(2**(s-s1))*(2**s1)-1)::ind
s2=s-s1
x1=0
```

```
y1=0
ind=0
do j1=2**s1-1,0,-1
        do ii=0,2**s-1
                    a=ii
                    do i1=0,s1-1,1
                        i0=site(i1)
                        call DTOBONEBIT(a,ia,i0,s)
                        site2(i1)=ia
                    enddo
                    call DTOB(j1,bin,s1)
                    oo=1
                    do k1=0,s1-1,1
                    oo=oo*(bin(k1)-site2(k1))
                    end do
                    if(abs(oo)==1) then
                        ind(x1)=ii
                            x1=x1+1
                    end if
            end do
end do
rdm=0.0d0
do t1=0,2**s1-1,1
   do t2=0,2**s1-1,1
      do z1=0,2**s2-1,1
         rdm(t1,t2)=rdm(t1,t2)+ &
            rho(ind((2**s2)*t1+z1),ind((2**s2)*t2+z1))
      enddo
   end do
end do
end subroutine
```

Example

In this example we are performing the partial trace on the four-qubit state $|\psi_2\rangle\langle\psi_2|$ and we will get the reduced density matrix ρ_{24} as an output by tracing out qubits 1 and 3.

```
program example
implicit none
integer,parameter::s=4,s1=2
integer::i,site(0:s1-1),j
real*8::psir(0:2**s-1),rhor(0:2**s-1,0:2**s-1)
real*8::rdm(0:2**s1-1,0:2**s1-1)
do i=0,2**s-1
    psir(i)=i+1
end do
call NORMALIZEVR(psir,2**s)
do i=0,2**s-1
    do j=0,2**s-1
        rhor(i,j)=psir(i)*psir(j)
    end do
```

```
end do
site(0)=1
site(1)=3
call PTRDMR(s,s1,site,rhor,rdm)
do i=0,2**s1-1
   write(111,*)real(rdm(i,:))
end do
end program
```

The file "fort.111" contains the entries of the required reduced density matrix ρ_{24}.

```
0.141711235  0.157754004  0.205882356  0.221925139
0.157754004  0.176470593  0.232620314  0.251336902
0.205882356  0.232620314  0.312834233  0.339572191
0.221925139  0.251336902  0.339572191  0.368983954
```

5.1.5   For a complex density matrix

```
subroutine PTRDMC(s,s1,site,rho,rdm)
```

Parameters

In/Out	Argument	Description
[in]	s	s (integer) is the total number of qubits
[in]	s1	s1 (integer) is the total number of qubits not to be traced out
[in]	site	site is an integer array of dimension $(0{:}s1-1)$ which contains the site indices of the qubits not to be traced out
[in]	rho	rho is a complex*16 array of dimension $(0{:}2^{**}s-1,0{:}2^{**}s-1)$ which is the input density matrix
[out]	rdm	rdm is a complex*16 array of dimension $(0{:}2^{**}s1-1,0{:}2^{**}s1-1)$ which is the required reduced density matrix

Implementation

```
subroutine PTRDMC(s,s1,site,rho,rdm)
implicit none
integer::s, s1,s2
real*8::trace
complex*16,dimension(0:2**s-1,0:2**s-1)::rho
complex*16,dimension(0:2**s1-1,0:2**s1-1)::rdm
integer::ii,i1,a,i0,ia,j1,oo,k1,x1,y1,j,t1,t2,z1,i
integer,dimension(0:s1-1)::site,site2
```

```
integer,dimension(0:s1-1)::bin
integer,dimension(0:(2**(s-s1))*(2**s1)-1)::ind
s2=s-s1
x1=0
y1=0
ind=0

do j1=2**s1-1,0,-1
        do ii=0,2**s-1
                a=ii
                do i1=0,s1-1,1
                    i0=site(i1)
                    call DTOBONEBIT(a,ia,i0,s)
                    site2(i1)=ia
                enddo
                call DTOB(j1,bin,s1)
                oo=1
                do k1=0,s1-1,1
                oo=oo*(bin(k1)-site2(k1))
                end do
                if(abs(oo)==1) then
                    ind(x1)=ii
                        x1=x1+1
                end if
        end do
end do
rdm=0.0d0
do t1=0,2**s1-1,1
   do t2=0,2**s1-1,1
     do z1=0,2**s2-1,1
        rdm(t1,t2)=rdm(t1,t2)+ &
             rho(ind((2**s2)*t1+z1),ind((2**s2)*t2+z1))
     enddo
   end do
end do
end subroutine
```

Example

In this example we are performing the partial trace on a four-qubit state $|\psi_3\rangle\langle\psi_3|$ and we will get the reduced density matrix ρ_4 as an output by tracing over qubits 1, 2 and 3.

```
program example
implicit none
integer,parameter::s=4,s1=1
integer::i,site(0:0),j
complex*16::psic(0:2**s-1),rhoc(0:2**s-1,0:2**s-1)
complex*16::rdm(0:2**s1-1,0:2**s1-1)
do i=0,2**s-1
    psic(i)=dcmplx(i,i+1)
end do
```

```
call NORMALIZEVC(psic,2**s)
do i=0,2**s-1
    do j=0,2**s-1
        rhoc(i,j)=psic(i)*conjg(psic(j))
    end do
end do
site(0)=3
call PTRDMC(s,s1,site,rhoc,rdm)
do i=0,2**s1-1
    do j=0,2**s1-1
        write(111,*)rdm(i,j)
    end do
end do
end program
```

The file "fort.111" contains the entries of the required reduced density matrix ρ_4.

(0.45321637426900580,0.0000000000000000)
(0.49707602339181289,2.92397660818712437E-003)
(0.49707602339181289,-2.92397660818712437E-003)
(0.54678362573099415,0.0000000000000000)

5.2 Partial transpose

Partial transpose [69, 73] is a very important operation in checking the separability of a given density matrix. Let us take an example of a bipartite composite system in the Hilbert space $H = H_A \otimes H_B$ and the density matrix corresponding to the composite system is ρ_{AB}. We can write the matrix elements of ρ_{AB} as

$$\rho_{AB} = \sum_{ijkl} r_{ik;jl}|i,j\rangle\langle k,l| \equiv \sum_{ijkl} r_{ik;jl}(|i\rangle\langle k|)_A \otimes (|j\rangle\langle l|)_B, \tag{5.5}$$

In the above equation, indices i,k are basis states in the Hilbert space H_A of subsystem A and j,l are basis states in the Hilbert space H_B of subsystem B. Here $r_{ik;jl}$ are the matrix elements. The partial transpose of ρ_{AB} with respect to the system B is given by,

$$\rho_{AB}^{T_B} = \sum_{ijkl} r_{ik;jl}(|i\rangle\langle k|)_A \otimes (|l\rangle\langle j|)_B. \tag{5.6}$$

Of course, $\rho_{AB}^{T_A}$ also can be found easily using the above. The recipes we have given can do partial transpose for any random partition of a bipartite multi-qubit system. Note that the flowchart for doing the partial transpose of real and complex density matrices are the same.

5.2.1 For a real density matrix

```
subroutine PTDMR(s,s1,site,rho,rhotp)
```

Parameters

In/Out	Argument	Description
[in]	s	s (integer) is the total number of qubits
[in]	s1	s1 (integer) is the total number of qubits to be transposed
[in]	site	site is an integer array of dimension $(0{:}s1-1)$ which contains the site indices of the qubits to be transposed
[in]	rho	rho is a real*8 array of dimension $(0{:}2^{**}s-1, 0{:}2^{**}s-1)$ which is the input density matrix
[out]	rhotp	rhotp is a real*8 array of dimension $(0{:}2^{**}s-1, 0{:}2^{**}s-1)$ which is the required partially transposed matrix

Implementation

```
subroutine PTDMR(s,s1,site,rho,rhotp)
implicit none
integer::s,i1,j1,i,j,oo,f,flag,t2,t3,t1,x1,s1
real*8::t4,t5,trace
integer,dimension(0:s1-1)::site
real*8,dimension(0:2**s-1,0:2**s-1):: rho,rhotp,rhot
integer,dimension(0:2**s-1)::ind1,ind2
integer,dimension(0:s-1)::m1,m2,m4,m5
integer,dimension(0:2*s-1)::m3
rhotp=rho
ind1=0
ind2=0
oo=0
do i=0,2**s-1,1
do j=0,2**s-1,1
f=0
    do j1=0,2**s-1
        if ( i==ind1(j1) .and. j==ind2(j1) ) then
            flag=1
        else
            flag=0
        end if
     f=f+flag
    end do

if(f==0 .and. i .ne. j) then
    call DTOB(i,m1,s)
    call DTOB(j,m2,s)

    do i1=0,2*s/2-1,1
      m3(i1)=m1(i1)
    end do
```

```fortran
   do i1=0,2*s/2-1,1
     m3(2*s/2+i1)=m2(i1)
   end do

   do x1=0,s1-1
     t1=m3(site(x1))
     m3(site(x1))=m3(site(x1)+s)
     m3(site(x1)+s)=t1
   end do

   do i1=0,2*s/2-1,1
     m4(i1)=m3(i1)
   end do
   do i1=0,2*s/2-1,1
     m5(i1)=m3(2*s/2+i1)
   end do

call BTOD(m4,t2,s)
call BTOD(m5,t3,s)

  t4=rho(t2,t3)
  t5=rho(i,j)
  rhotp(i,j)=t4
  rhotp(t2,t3)=t5
  ind1(oo)=t2
  ind2(oo)=t3

end if

end do
oo=oo+1
end do
end subroutine
```

Example

In this example we are performing the partial transpose over the second qubit on the two-qubit density matrix $|\psi_2\rangle\langle\psi_2|$ and we will get ρ^{T_2} as the output.

```fortran
program example
implicit none
integer,parameter::s=2,s1=1
integer::i,site(0:0),j
real*8::psir(0:2**s-1),rhor(0:2**s-1,0:2**s-1)
real*8::rhotp(0:2**s-1,0:2**s-1)
do i=0,2**s-1
    psir(i)=i+1
end do
call NORMALIZEVR(psir,2**s)
do i=0,2**s-1
```

```
    do j=0,2**s-1
        rhor(i,j)=psir(i)*psir(j)
    end do
end do
site(0)=1
call PTDMR(s,s1,site,rhor,rhotp)
do i=0,2**s-1
    write(111,*)real(rhotp(i,:))
end do
end program
```

The file "fort.111" contains the entries of the required partially transposed matrix.

3.33333351E-02	6.66666701E-02	0.100000001	0.200000003
6.66666701E-02	0.133333340	0.133333340	0.266666681
0.100000001	0.133333340	0.300000012	0.400000006
0.200000003	0.266666681	0.400000006	0.533333361

5.2.2 For a complex density matrix

```
subroutine PTDMC(s,s1,site,rho,rhotp)
```

Parameters

In/Out	Argument	Description
[in]	s	s (integer) is the total number of qubits
[in]	s1	s1 (integer) is the total number of qubits to be transposed
[in]	site	site is an integer array of dimension $(0:s1-1)$ which contains the site indices of the qubits to be transposed
[in]	rho	rho is a complex*16 array of dimension $(0:2**s-1,0:2**s-1)$ which is the input density matrix
[out]	rhotp	rhotp is a complex*16 array of dimension $(0:2**s-1,0:2**s-1)$ which is the required partially transposed matrix

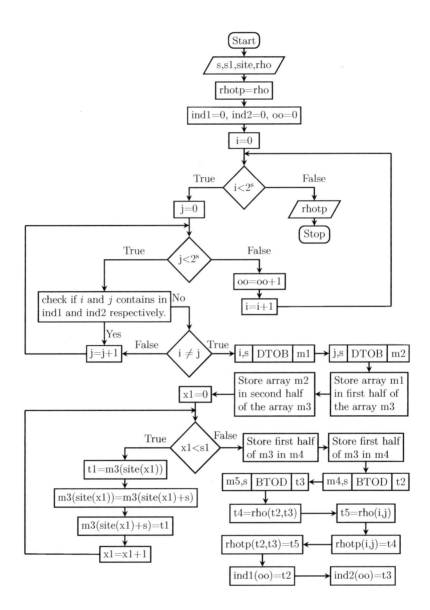

FIGURE 5.5: Flowchart of PTDMR and PTDMC

Implementation

```
subroutine PTDMC(s,s1,site,rho,rhotp)
implicit none
integer::s,i1,j1,i,j,oo,f,flag,t2,t3,t1,x1,s1
complex*16::t4,t5
real*8::trace
integer,dimension(0:s1-1)::site
complex*16,dimension(0:2**s-1,0:2**s-1):: rho,rhotp
integer,dimension(0:2**s-1)::ind1,ind2
integer,dimension(0:s-1)::m1,m2,m4,m5
integer,dimension(0:2*s-1)::m3
rhotp=rho
ind1=0
ind2=0
oo=0
do i=0,2**s-1,1
do j=0,2**s-1,1
      f=0
   do j1=0,2**s-1
       if ( i==ind1(j1) .and. j==ind2(j1) ) then
            flag=1
       else
            flag=0
       end if
     f=f+flag
   end do

if(f==0 .and. i .ne. j) then
    call DTOB(i,m1,s)
    call DTOB(j,m2,s)

   do i1=0,2*s/2-1,1
      m3(i1)=m1(i1)
   end do

   do i1=0,2*s/2-1,1
      m3(2*s/2+i1)=m2(i1)
   end do

   do x1=0,s1-1
      t1=m3(site(x1))
      m3(site(x1))=m3(site(x1)+s)
      m3(site(x1)+s)=t1
   end do

   do i1=0,2*s/2-1,1
      m4(i1)=m3(i1)
   end do
   do i1=0,2*s/2-1,1
      m5(i1)=m3(2*s/2+i1)
```

```
      end do

call BTOD(m4,t2,s)
call BTOD(m5,t3,s)

  t4=rho(t2,t3)
  t5=rho(i,j)
  rhotp(i,j)=t4
  rhotp(t2,t3)=t5
  ind1(oo)=t2
  ind2(oo)=t3

end if

end do
oo=oo+1
end do
end subroutine
```

Example

In this example we are performing the partial transpose over the second qubit on the two-qubit density matrix $|\psi_3\rangle\langle\psi_3|$ and we will get ρ^{T_2} as the output.

```
program example
implicit none
integer,parameter::s=2,s1=1
integer::i,site(0:0),j
complex*16::psic(0:2**s-1),rhoc(0:2**s-1,0:2**s-1)
complex*16::rhotp(0:2**s-1,0:2**s-1)
do i=0,2**s-1
    psic(i)=dcmplx(i,i+1)
end do
call NORMALIZEVC(psic,2**s)
do i=0,2**s-1
    do j=0,2**s-1
        rhoc(i,j)=psic(i)*conjg(psic(j))
    end do
end do
site(0)=1
call PTDMC(s,s1,site,rhoc,rhotp)
do i=0,2**s-1
    do j=0,2**s-1
        write(111,*)rhotp(i,j)
    end do
end do
end program
```

The file "fort.111" contains the entries of the required partially transposed matrix.

```
      (2.27272727272727279E-002,0.0000000000000000)
  (4.54545454545454558E-002,-2.27272727272727279E-002)
```

```
(6.81818181818181906E-002,4.54545454545454558E-002)
      (0.18181818181818182,2.27272727272727210E-002)
(4.54545454545454558E-002,2.27272727272727279E-002)
          (0.11363636363636365,0.0000000000000000)
(9.09090909090909116E-002,6.81818181818181906E-002)
      (0.25000000000000000,4.54545454545454697E-002)
(6.81818181818181906E-002,-4.54545454545454558E-002)
(9.09090909090909116E-002,-6.81818181818181906E-002)
          (0.29545454545454547,0.0000000000000000)
      (0.40909090909090917,-2.27272727272727348E-002)
      (0.18181818181818182,-2.27272727272727210E-002)
      (0.25000000000000000,-4.54545454545454697E-002)
      (0.40909090909090917,2.27272727272727348E-002)
          (0.56818181818181823,0.0000000000000000)
```

5.2.3 For a real state

`subroutine PTVR(s,s1,site,vin,rhotp)`

Parameters

In/Out	Argument	Description
[in]	s	s (integer) is the total number of qubits
[in]	s1	s1 (integer) is the total number of qubits to be transposed
[in]	site	site is an integer array of dimension $(0:s1-1)$ which contains the site indices of the qubits to be transposed
[in]	vin	vin is a real*8 array of dimension $(0:2**s-1)$ which is the input state
[out]	rhotp	rhotp is a real*8 array of dimension $(0:2**s-1,0:2**s-1)$ which is the required partially transposed matrix

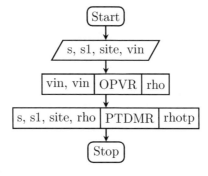

FIGURE 5.6: Flowchart of PTVR

Implementation

```
subroutine PTVR(s,s1,site,vin,rhotp)
implicit none
integer::s,s1,i,j
integer,dimension(0:s1-1)::site
real*8,dimension(0:2**s-1)::vin
real*8,dimension(0:2**s-1,0:2**s-1):: rho,rhotp
call OPVR(vin,vin,2**s,2**s,rho)
call PTDMR(s,s1,site,rho,rhotp)
end subroutine
```

Example

In this example we are performing the partial transpose over the second qubit on the two-qubit pure state $|\psi_2\rangle$ and we will get ρ^{T_2} as the output.

```
program example
implicit none
integer,parameter::s=2,s1=1
integer::i,site(0:0)
real*8::psir(0:2**s-1),rhotp(0:2**s-1,0:2**s-1)
do i=0,2**s-1
    psir(i)=i+1
end do
call NORMALIZEVR(psir,2**s)
site(0)=1
call PTVR(s,s1,site,psir,rhotp)
do i=0,2**s-1
    write(111,*)real(rhotp(i,:))
end do
end program
```

The file "fort.111" contains the entries of the required partially transposed matrix.

3.33333351E-02	6.66666701E-02	0.100000001	0.200000003
6.66666701E-02	0.133333340	0.133333340	0.266666681
0.100000001	0.133333340	0.300000012	0.400000006
0.200000003	0.266666681	0.400000006	0.533333361

5.2.4 For a complex state

```
subroutine PTVC(s,s1,site,vin,rhotp)
```

Parameters

In/Out	Argument	Description
[in]	s	s (integer) is the total number of qubits
[in]	s1	s1 (integer) is the total number of qubits to be transposed
[in]	site	site is an integer array of dimension $(0:s1-1)$ which contains the site indices of the qubits to be transposed
[in]	vin	vin is a complex*16 array of dimension $(0:2**s-1)$ which is the input state
[out]	rhotp	rhotp is a complex*16 array of dimension $(0:2**s-1,0:2**s-1)$ which is the required partially transposed matrix

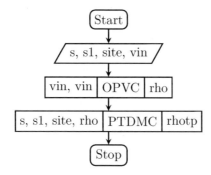

FIGURE 5.7: Flowchart of PTVC

Implementation

```
subroutine PTVC(s,s1,site,vin,rhotp)
implicit none
integer::s,s1,i,j
integer,dimension(0:s1-1)::site
complex*16,dimension(0:2**s-1,0:2**s-1):: rho,rhotp
complex*16,dimension(0:2**s-1)::vin
call OPVC(vin,vin,2**s,2**s,rho)
call PTDMC(s,s1,site,rho,rhotp)
end subroutine
```

Example

In this example we are performing the partial transpose over the second qubit on the two-qubit pure state $|\psi_3\rangle$ and we will get ρ^{T_2} as the output.

```
program example
implicit none
```

```
integer,parameter::s=2,s1=1
integer::i,site(0:0),j
complex*16::psic(0:2**s-1),rhotp(0:2**s-1,0:2**s-1)
real*8::be
do i=0,2**s-1
   psic(i)=dcmplx(i,i+1)
end do
call NORMALIZEVC(psic,2**s)
site(0)=1
call PTVC(s,s1,site,psic,rhotp)
do i=0,2**s-1
   do j=0,2**s-1
      write(111,*)rhotp(i,j)
   end do
end do
end program
```

The file "fort.111" contains the entries of required partially transposed matrix.

```
        (2.27272727272727279E-002,0.0000000000000000)
     (4.54545454545454558E-002,-2.27272727272727279E-002)
     (6.81818181818181906E-002,4.54545454545454558E-002)
        (0.18181818181818182,2.27272727272727210E-002)
     (4.54545454545454558E-002,2.27272727272727279E-002)
           (0.11363636363636365,0.0000000000000000)
     (9.09090909090909116E-002,6.81818181818181906E-002)
        (0.25000000000000000,4.54545454545454697E-002)
     (6.81818181818181906E-002,-4.54545454545454558E-002)
     (9.09090909090909116E-002,-6.81818181818181906E-002)
           (0.29545454545454547,0.0000000000000000)
        (0.40909090909090917,-2.27272727272727348E-002)
        (0.18181818181818182,-2.27272727272727210E-002)
        (0.25000000000000000,-4.54545454545454697E-002)
        (0.40909090909090917,2.27272727272727348E-002)
           (0.56818181818181823,0.0000000000000000)
```

5.3   Concurrence

Concurrence [74,75] is a measure of entanglement defined for both pure and mixed state of two qubits and it is given as,

$$C_{ij} = \text{Max}\ (0, \sqrt{\lambda_1} - \sqrt{\lambda_2} - \sqrt{\lambda_3} - \sqrt{\lambda_4}). \tag{5.7}$$

Where λ_i's are the eigenvalues of the matrix $\rho_{ij}\tilde{\rho}_{ij}$ in the non-increasing order. Here ρ_{ij} is the reduced density matrix corresponding to the qubits i and j between which we are interested in calculating the concurrence. $\tilde{\rho}_{ij}$ is defined as,

$$\tilde{\rho}_{ij} = (\sigma_y \otimes \sigma_y)\rho_{ij}^*(\sigma_y \otimes \sigma_y). \tag{5.8}$$

Note carefully that, if we have any pure or mixed state of N qubits, then once we have the reduced density matrix between any two qubits i and j as ρ_{ij}, we can directly use the above prescription to calculate concurrence. It is worth mentioning that, it is a measure of entanglement that has a closed form expression for two-qubit states and at the same time cannot be generalized to higher dimensions. Note that, in the recipes given, we can calculate the concurrence of any pair ij $(i < j)$ of a N-qubit state or a density matrix.

5.3.1 For a real state

```
subroutine CONVR(s,i0,j0,vin,con)
```

Parameters

In/Out	Argument	Description
[in]	s	s (integer) is the total number of qubits
[in]	i0, j0	i0 (integer) and j0 (integer) are the site indices of the qubits (such that i0 < j0) between which we require the concurrence
[in]	vin	vin is a real*8 array of dimension (0:2**s−1) which is the input state
[out]	con	con (real*8) is the value of concurrence

Implementation

```
subroutine CONVR(s,i0,j0,vin,con)
implicit none
integer::s,i0,j0 ,site(0:1),i
real*8::vin(0:2**s-1),evs
real*8,dimension(0:3,0:3)::rdm,MM1,MM2,MM3,MM4
INTEGER::INFO, LDA, LWORK, NN,LDVL,LDVR
parameter(NN=4,LDA=NN,LWORK=4*NN,LDVL=NN,LDVR=NN)
real*8::WORK(0:LWORK-1),AA2(0:LDA-1,0:NN-1)
real*8::VL(0:NN-1,0:NN-1),VR(0:NN-1,0:NN-1)
real*8::WR(0:NN-1),WI(0:NN-1)
real*8::ev(NN) ,con,c2,x11,const,sigmay(0:3,0:3)
sigmay(0,:)=(/0.0,0.0,0.0,-1.0/)
sigmay(1,:)=(/0.0,0.0,1.0,0.0/)
sigmay(2,:)=(/0.0,1.0,0.0,0.0/)
sigmay(3,:)=(/-1.0,0.0,0.0,0.0/)
site(0)=i0
site(1)=j0
call PTRVR(s,2,site,vin,rdm)
MM1=matmul(rdm,sigmay)
MM2=matmul(rdm,sigmay)
MM3=matmul(MM1,MM2)
call dgeev('N', 'N', NN, MM3, LDA, WR, &
```

```
            WI, VL, LDVL, VR, LDVR, WORK, LWORK, INFO)
do i=1,NN,1
    if (WR(i-1) .lt. 0.0000000000001) then
        WR(i-1)=0.0d0
    end if
    ev(i)=WR(i-1)
end do

do i=2,4
    if(ev(i)>=ev(1))then
        evs=ev(i)
        ev(i)=ev(1)
        ev(1)=evs
    end if
end do

do i=3,4
    if(ev(i)>=ev(2))then
        evs=ev(i)
        ev(i)=ev(2)
        ev(2)=evs
    end if
end do

do i=4,4
    if(ev(i)>=ev(3))then
        evs=ev(i)
        ev(i)=ev(3)
        ev(3)=evs
    end if
end do
c2=0
con=0
c2=dsqrt(ev(1))-dsqrt(ev(2))-dsqrt(ev(3))-dsqrt(ev(4))
if(c2>0)then
    con=c2
else
    con=0
end if
end subroutine
```

Example

In this example we are finding the value of concurrence C_{35} of a six-qubit state $|\psi_2\rangle$.

```
program example
implicit none
integer,parameter::s=6
integer::i
real*8::psir(0:2**s-1),con
do i=0,2**s-1
```

```
    psir(i)=i+1
end do
call NORMALIZEVR(psir,2**s)
call CONVR(s,2,4,psir,con)
write(111,*)con
end program
```

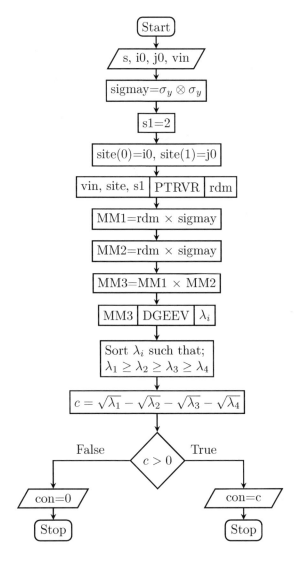

FIGURE 5.8: Flowchart of CONVR

The file "fort.111" contains the required value of concurrence C_{35}.

```
5.7245080501163656E-003
```

5.3.2   For a complex state

```
subroutine CONVC(s,i0,j0,vin,con)
```

Parameters

In/Out	Argument	Description
[in]	s	s (integer) is the total number of qubits
[in]	i0, j0	i0 (integer) and j0 (integer) are the site indices of the qubits (such that i0 < j0) between which we require the concurrence
[in]	vin	vin is a complex*16 array of dimension $(0:2^{**}s-1)$ which is the input state
[out]	con	con (real*8) is the value of concurrence

Implementation

```
subroutine CONVC(s,i0,j0,vin,con)
implicit none
integer::s,i0,j0 ,site(0:1),i
complex*16::vin(0:2**s-1),evs
complex*16,dimension(0:3,0:3)::rdm,MM1,MM2,MM3
INTEGER::INFO, LDA, LWORK, N,LDVL,LDVR
parameter(N=4,LDA=N,LWORK=2*N,LDVL=N,LDVR=N)
complex*16::WORK(0:LWORK-1),VL(0:N-1,0:N-1)
complex*16::VR(0:N-1,0:N-1),W(0:N-1)
real*8::RWORK(0:2*N-1),ev(N) ,con,c2,x11,sigmay(0:3,0:3)
sigmay(0,:)=(/0.0,0.0,0.0,-1.0/)
sigmay(1,:)=(/0.0,0.0,1.0,0.0/)
sigmay(2,:)=(/0.0,1.0,0.0,0.0/)
sigmay(3,:)=(/-1.0,0.0,0.0,0.0/)
site(0)=i0
site(1)=j0
call PTRVC(s,2,site,vin,rdm)
MM1=matmul(rdm,sigmay)
MM2=matmul(conjg(rdm),sigmay)
MM3=matmul(MM1,MM2)
call zgeev('N', 'N', N, MM3, LDA, W, VL, LDVL, &
          VR, LDVR, WORK, LWORK, RWORK, INFO)
do i=1,N,1
   if (real(W(i-1))<0.0000000000001) then
      W(i-1)=0.0d0
   end if
   ev(i)=W(i-1)
end do
```

```
do i=2,4
    if(ev(i)>=ev(1))then
        evs=ev(i)
        ev(i)=ev(1)
        ev(1)=evs
    end if
end do

do i=3,4
    if(ev(i)>=ev(2))then
        evs=ev(i)
        ev(i)=ev(2)
        ev(2)=evs
    end if
end do

do i=4,4
    if(ev(i)>=ev(3))then
        evs=ev(i)
        ev(i)=ev(3)
        ev(3)=evs
    end if
end do
c2=0
con=0
c2=dsqrt(ev(1))-dsqrt(ev(2))-dsqrt(ev(3))-dsqrt(ev(4))
if(c2>0)then
    con=c2
else
    con=0
end if
end subroutine
```

Example

In this example we are finding the value of concurrence C_{36} of the six-qubit state $|\psi_3\rangle$.

```
program example
implicit none
integer,parameter::s=6
integer::i
complex*16::psic(0:2**s-1)
real*8::con
do i=0,2**s-1
    psic(i)=dcmplx(i,i+1)
end do
call NORMALIZEVC(psic,2**s)
call CONVC(s,2,5,psic,con)
write(111,*)con
end program
```

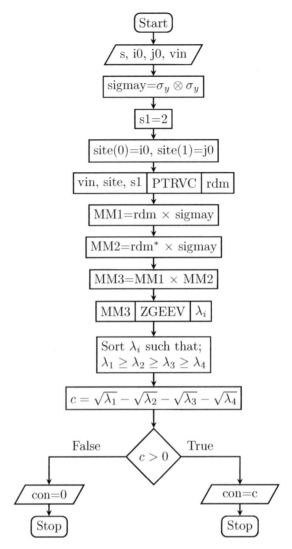

FIGURE 5.9: Flowchart of CONVC

The file "fort.111" contains the required value of concurrence C_{36}.

2.9293299161676856E-003

5.3.3 For a real density matrix

```
subroutine CONDMR(s,i0,j0,rho,con)
```

Parameters

In/Out	Argument	Description
[in]	s	s (integer) is the total number of qubits
[in]	i0, j0	i0 (integer) and j0 (integer) are the site indices of the qubits (such that i0 < j0) between which we require the concurrence
[in]	rho	rho is a real*8 array of dimension $(0:2^{**}s-1,0:2^{**}s-1)$ which is the input density matrix
[out]	con	con (real*8) is the value of concurrence

Implementation

```fortran
subroutine CONDMR(s,i0,j0,rho,con)
implicit none
integer::s,i0,j0 ,site(0:1),i
real*8::rho(0:2**s-1,0:2**s-1)
real*8,dimension(0:3,0:3)::rdm,MM1,MM2,MM3,MM4
INTEGER::INFO, LDA, LWORK, N,LDVL,LDVR
parameter(N=4,LDA=N,LWORK=4*N,LDVL=N,LDVR=N)
real*8::WORK(0:LWORK-1),VL(0:N-1,0:N-1)
real*8::VR(0:N-1,0:N-1),WR(0:N-1),WI(0:N-1)
real*8::RWORK(0:2*N-1),ev(N),con,c2,sigmay(0:3,0:3),evs
sigmay(0,:)=(/0.0,0.0,0.0,-1.0/)
sigmay(1,:)=(/0.0,0.0,1.0,0.0/)
sigmay(2,:)=(/0.0,1.0,0.0,0.0/)
sigmay(3,:)=(/-1.0,0.0,0.0,0.0/)
site(0)=i0
site(1)=j0
call PTRDMR(s,2,site,rho,rdm)
MM1=matmul(rdm,sigmay)
MM2=matmul(rdm,sigmay)
MM3=matmul(MM1,MM2)
call dgeev('N', 'N', N, MM3, LDA, WR, WI, VL, &
                LDVL, VR, LDVR, WORK, LWORK, INFO)
do i=1,N,1
   if (WR(i-1)<0.0000000000001) then
       WR(i-1)=0.0d0
   end if
   ev(i)=WR(i-1)
end do

do i=2,4
   if(ev(i)>=ev(1))then
       evs=ev(i)
       ev(i)=ev(1)
       ev(1)=evs
```

```
      end if
end do

do i=3,4
    if(ev(i)>=ev(2))then
        evs=ev(i)
        ev(i)=ev(2)
        ev(2)=evs
    end if
end do

do i=4,4
    if(ev(i)>=ev(3))then
        evs=ev(i)
        ev(i)=ev(3)
        ev(3)=evs
    end if
end do
c2=0
con=0
c2=dsqrt(ev(1))-dsqrt(ev(2))-dsqrt(ev(3))-dsqrt(ev(4))
if(c2>0)then
    con=c2
else
    con=0
end if
end subroutine
```

Example

In this example we are finding the value of the concurrence C_{36} of a six-qubit density matrix $|\psi_2\rangle\langle\psi_2|$.

```
program example
implicit none
integer,parameter::s=6
integer::i,j
real*8::psir(0:2**s-1),con,rhor(0:2**s-1,0:2**s-1)
do i=0,2**s-1
    psir(i)=i+1
end do
call NORMALIZEVR(psir,2**s)
do i=0,2**s-1
    do j=0,2**s-1
        rhor(i,j)=psir(i)*psir(j)
    end do
end do
call CONDMR(s,2,5,rhor,con)
write(111,*)con
end program
```

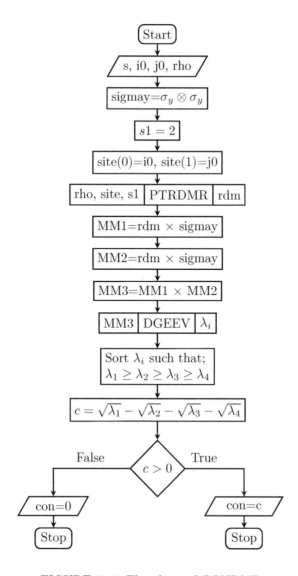

FIGURE 5.10: Flowchart of CONDMR

The file "fort.111" contains the required value of concurrence C_{36}.

2.8622540248506665E-003

5.3.4 For a complex density matrix

```
subroutine CONDMC(s,i0,j0,rho,con)
```

Parameters

In/Out	Argument	Description
[in]	s	s (integer) is the total number of qubits
[in]	i0, j0	i0 (integer) and j0 (integer) are the site indices of the qubits (such that i0 < j0) between which we require the concurrence
[in]	rho	rho is a complex*16 array of dimension $(0:2^{**}s-1,0:2^{**}s-1)$ which is the input density matrix
[out]	con	con (real*8) is the value of concurrence

Implementation

```
subroutine CONDMC(s,i0,j0,rho,con)
implicit none
integer::s,i0,j0 ,site(0:1),i
complex*16::rho(0:2**s-1,0:2**s-1)
complex*16,dimension(0:3,0:3)::rdm,MM1,MM2,MM3,MM4
INTEGER::INFO, LDA, LWORK, N,LDVL,LDVR
parameter(N=4,LDA=N,LWORK=2*N,LDVL=N,LDVR=N)
complex*16::WORK(0:0,0:LWORK-1),AA2(0:LDA-1,0:N-1)
complex*16::VL(0:N-1,0:N-1),VR(0:N-1,0:N-1),W(0:N-1)
real*8::RWORK(0:2*N-1),ev(N) ,con,c2,x11,const
real*8::sigmay(0:3,0:3),evs
sigmay(0,:)=(/0.0,0.0,0.0,-1.0/)
sigmay(1,:)=(/0.0,0.0,1.0,0.0/)
sigmay(2,:)=(/0.0,1.0,0.0,0.0/)
sigmay(3,:)=(/-1.0,0.0,0.0,0.0/)
site(0)=i0
site(1)=j0
call PTRDMC(s,2,site,rho,rdm)
MM1=matmul(rdm,sigmay)
MM2=matmul(conjg(rdm),sigmay)
MM3=matmul(MM1,MM2)
call zgeev('N', 'N', N, MM3, LDA, W, VL, LDVL, &
              VR, LDVR, WORK, LWORK, RWORK, INFO)
do i=1,N,1
   if (real(W(i-1))<0.0000000000001) then
      W(i-1)=0.0d0
   end if
   ev(i)=W(i-1)
end do

do i=2,4
   if(ev(i)>=ev(1))then
      evs=ev(i)
```

```
        ev(i)=ev(1)
        ev(1)=evs
    end if
end do

do i=3,4
    if(ev(i)>=ev(2))then
        evs=ev(i)
        ev(i)=ev(2)
        ev(2)=evs
    end if
end do

do i=4,4
    if(ev(i)>=ev(3))then
        evs=ev(i)
        ev(i)=ev(3)
        ev(3)=evs
    end if
end do
c2=0
con=0
c2=dsqrt(ev(1))-dsqrt(ev(2))-dsqrt(ev(3))-dsqrt(ev(4))
if(c2>0)then
    con=c2
else
    con=0
end if
end subroutine
```

Example

In this example we are finding the value of the concurrence C_{36} of the six-qubit density matrix $|\psi_3\rangle\langle\psi_3|$.

```
program example
implicit none
integer,parameter::s=6
integer::i,j
complex*16::psic(0:2**s-1),rhoc(0:2**S-1,0:2**s-1)
real*8::con
do i=0,2**s-1
    psic(i)=dcmplx(i,i+1)
end do
call NORMALIZEVC(psic,2**s)
do i=0,2**s-1
    do j=0,2**s-1
        rhoc(i,j)=psic(i)*conjg(psic(j))
    end do
end do
call CONDMC(s,2,5,rhoc,con)
```

```
write(111,*)con
end program
```

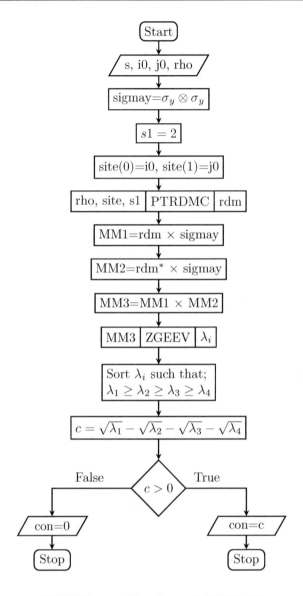

FIGURE 5.11: Flowchart of CONDMC

The file "fort.111" contains the required value of concurrence C_{36}.

```
2.9293299161676856E-003
```

There are different variants of concurrence in the entanglement literature as follows. *I*-concurrence [76] of any multi-qubit pure state is defined as:

$$C_I = \sqrt{2(1 - Tr(\rho_A^2))}. \tag{5.9}$$

Where ρ_A is the reduced density matrix of partition A. The example of finding the *I*-concurrence for state $(|00\rangle + |11\rangle)/\sqrt{2}$ is given below:

```
program example
integer,parameter::s=2
real*8::psi(0:3),rdm(0:1,0:1),ci
complex*16::trace
integer::site(0:0)
call NQBGHZ(s,psi)
site(0)=0
call PTRVR(s,1,site,psi,rdm)
call TRPOWNSR(rdm,2,2.0d0,trace)
 ci=dsqrt(2*(1-real(trace)))
print*,ci
end program
```

which prints the I-concurrence equal to 1.

There exists another type of concurrence called the N-concurrence [77] which is defined as,

$$C_N = 2^{1-\frac{N}{2}} \sqrt{2^N - 2 - \sum_i Tr(\rho_i^2)}. \qquad (5.10)$$

The example of finding the N-concurrence for state $(|00\rangle + |11\rangle)/\sqrt{2}$ is given below:

```
program example
integer,parameter::s=2
real*8::psi(0:3),rdm(0:1,0:1),cn,summ
complex*16::trace
integer::site(0:0),i
call NQBGHZ(s,psi)
summ=0.0d0
do i=0,s-1
    site(0)=i
    call PTRVR(s,1,site,psi,rdm)
    call TRPOWNSR(rdm,2,2.0d0,trace)
    summ=summ+trace
end do
 cn=(2**(1-s/2))*sqrt(2**s-s-summ)
print*,cn
end program
```

which prints the N-concurrence equal to 1.

5.4   Block entropy

For a two-qubit pure state the von Neumann entropy [11, 78] is an entanglement measure which is also called the entanglement entropy. If we have a two-qubit state in Hilbert space $H = H_A \otimes H_B$ where A represents the first qubit and B represent the second qubit. Now the state of the composite system of two qubits can be written as ρ_{AB} which is a 4×4 matrix. To know the entanglement content between A and B we calculate the von Neumann entropy of any of the subsystem reduced density matrices ρ_A or ρ_B which are 2×2 matrices.

Then the entanglement entropy is defined as,

$$S(\rho_A) = S(\rho_B) = -\sum_{i=1}^{2} \lambda_i \log_2 \lambda_i. \tag{5.11}$$

where λ_i's are the non zero eigenvalues of the matrix ρ_A or ρ_B which will be equal due to Schmidt decomposition. A simple example will illustrate the point, consider any Bell state, they are pure states in two qubits, so the von Neumann entropy of a Bell state will be zero (as complete information means no ignorance about the state). However, to know the entanglement between the two qubits making the Bell state, if we calculate the subsystem von Neumann entropies as shown in the above equation, they turn out to be unity, meaning they are maximally entangled.

A very important extension of von Neumann entropy for a multi qubit pure state is the concept of block entropy [79], where we consider a N qubit pure state which we can split it into two parts namely, L and $N - L$ subsystem such that $L < N$. We now write the block entropy for the L and $N - L$ blocks as,

$$S(\rho_L) = -Tr(\rho_L \log_2 \rho_L) = -\sum_{i=1}^{dim(\rho_L)} \lambda_i' \log_2 \lambda_i' \tag{5.12}$$

$$S(\rho_{N-L}) = -Tr(\rho_{N-L} \log_2 \rho_{N-L}) = -\sum_{i=1}^{dim(\rho_{N-L})} \lambda_i'' \log_2 \lambda_i'' \tag{5.13}$$

Note that, in the above equations, λ_i''s and λ_i'''s are the non-zero eigenvalues of the matrix ρ_L and ρ_{N-L}, respectively which are equal. The block entropy quantifies the entanglement between two blocks, but there may be arbitrary number of qubits inside each block. Block entropy is a very important multi-qubit entanglement in many border line areas between condensed matter physics and quantum information like spin chains, etc.

5.4.1   For a real state

```
subroutine BERPVR(s,s1,site,vin,be)
```

Parameters

In/Out	Argument	Description
[in]	s	s (integer) is the total number of qubits
[in]	s1	s1 (integer) is the total number of qubits not to be traced out
[in]	site	site is an integer array of dimension (0:s1−1) which contains the site indices of the qubits not to be traced out
[in]	vin	vin is a real*8 array of dimension (0:2**s−1) which is the input state
[out]	be	be (real*8) is the value of block entropy

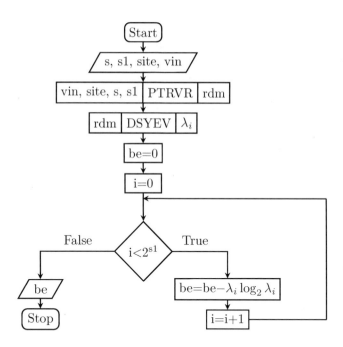

FIGURE 5.12: Flowchart of BERPVR

Implementation

```
subroutine BERPVR(s,s1,site,vin,be)
implicit none
integer::s, s1,site(0:s1-1)
real*8::be,betemp,vin(0:2**s-1),rdm(0:2**s1-1,0:2**s1-1)
integer::i,INFO
real*8::WORK(0:3*(2**s1)-1),W(0:2**s1-1)
call PTRVR(s,s1,site,vin,rdm)
call DSYEV('N','U', 2**s1, rdm, 2**s1, &
           W, WORK,3*(2**s1)-1,INFO)
be=0
betemp=0
do i=0,2**s1-1,1
    if (W(i)>0.0000000000001) then
        betemp=-(dlog(abs(W(i)))*abs(W(i)))/dlog(2.0d0)
        be=be+betemp
    end if
end do
end subroutine
```

Example

In this example we are finding the value of the block entropy between the blocks 135 and 246 of the six-qubit state $|\psi_2\rangle$.

```
program example
implicit none
```

```
integer,parameter::s=6
integer::i,site(0:2)
real*8::psir(0:2**s-1),be
do i=0,2**s-1
    psir(i)=i+1
end do
call NORMALIZEVR(psir,2**s)
site(0)=0
site(1)=2
site(2)=4
call BERPVR(s,3,site,psir,be)
write(111,*)be
end program
```

The file "fort.111" contains the required value of block entropy.

```
7.8350818181306514E-002
```


5.4.2   For a complex state

```
subroutine BERPVC(s,s1,site,vin,be)
```

Parameters

In/Out	Argument	Description
[in]	s	s (integer) is the total number of qubits
[in]	s1	s1 (integer) is the total number of qubits not to be traced out.
[in]	site	site is an integer array of dimension (0:s1−1) which contains the site indices of the qubits not to be traced out
[in]	vin	vin is a complex*16 array of dimension (0:2**s−1) which is the input state
[out]	be	be (real*8) is the value of block entropy

Implementation

```
subroutine BERPVC(s,s1,site,vin,be)
implicit none
integer::s, s1,site(0:s1-1)
real*8::be,betemp
complex*16::vin(0:2**s-1),rdm(0:2**s1-1,0:2**s1-1)
integer::i,INFOO
complex*16::WORKK(0:2*(2**s1)-1)
real*8::RWORKK(0:3*(2**s1)-3),WW(0:2**s1-1)
call PTRVC(s,s1,site,vin,rdm)
```

```
call zheev('N','U', 2**s1, rdm, 2**s1, &
           WW, WORKK,2*(2**s1)-1,RWORKK,INFOO)
be=0
betemp=0
do i=0,2**s1-1,1
    if (WW(i)>0.0000000000001) then
        betemp=-(dlog(abs(WW(i)))*abs(WW(i)))/dlog(2.0d0)
        be=be+betemp
    end if
end do
end subroutine
```

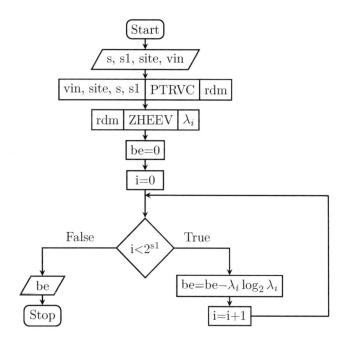

FIGURE 5.13: Flowchart of BERPVC

Example

In this example we are finding the value of block entropy between the blocks 135 and 246 of the six-qubit state $|\psi_3\rangle$.

```
program example
implicit none
integer,parameter::s=6
integer::i,site(0:2)
complex*16::psic(0:2**s-1)
real*8::be
do i=0,2**s-1
psic(i)=dcmplx(i,i+1)
end do
call NORMALIZEVC(psic,2**s)
site(0)=0
```

```
site(1)=2
site(2)=4
call BERPVC(s,3,site,psic,be)
write(111,*)be
end program
```

The file "fort.111" contains the required value of block entropy.

```
8.1419480965577343E-002
```

5.4.3   For a real density matrix

```
subroutine BERPDMR(s,s1,site,rho,be)
```

Parameters

In/Out	Argument	Description
[in]	s	s (integer) is the total number of qubits
[in]	s1	s1 (integer) is the total number of qubits not to be traced out
[in]	site	site is an integer array of dimension $(0:s1-1)$ which contains the site indices of the qubits not to be traced out
[in]	rho	rho is a real*8 array of dimension $(0:2^{**}s-1,0:2^{**}s-1)$ which is the input density matrix
[out]	be	be (real*8) is the value of block entropy

Implementation

```
subroutine BERPDMR(s,s1,site,rho,be)
implicit none
integer::s, s1,i,INFOO,site(0:s1-1)
real*8::trace,be,betemp,rho(0:2**s-1,0:2**s-1)
real*8::rdm(0:2**s1-1,0:2**s1-1)
real*8::WORKK(0:3*(2**s1)-1),WW(0:2**s1-1)
call PTRDMR(s,s1,site,rho,rdm)
call DSYEV('N','U', 2**s1, rdm, 2**s1, &
                 WW, WORKK,3*(2**s1)-1,INFOO)
be=0
betemp=0
do i=0,2**s1-1,1
    if (WW(i)>0.0000000000001) then
        betemp=-(dlog(abs(WW(i)))*abs(WW(i)))/dlog(2.0d0)
        be=be+betemp
    end if
end do
```

```
end subroutine
```

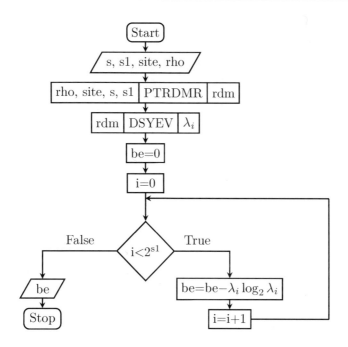

FIGURE 5.14: Flowchart of BERPDMR

Example

In this example we are finding the value of block entropy between the blocks 135 and 246 of the six-qubit density matrix $|\psi_2\rangle\langle\psi_2|$.

```
program example
implicit none
integer,parameter::s=6
integer::i,site(0:2),j
real*8::psir(0:2**s-1),be,rhor(0:2**s-1,0:2**s-1)
do i=0,2**s-1
    psir(i)=i+1
end do
call NORMALIZEVR(psir,2**s)
do i=0,2**s-1
    do j=0,2**s-1
        rhor(i,j)=psir(i)*psir(j)
    end do
end do
site(0)=1
site(1)=3
site(2)=5
call BERPDMR(s,3,site,rhor,be)
write(111,*)be
end program
```

The file "fort.111" contains the required value of block entropy.

7.8350818181306486E-002

5.4.4 For a complex density matrix

---
subroutine BERPDMC(s,s1,site,rho,be)

Parameters

In/Out	Argument	Description
[in]	s	s (integer) is the total number of qubits
[in]	s1	s1 (integer) is the total number of qubits not to be traced out
[in]	site	site is an integer array of dimension $(0{:}s1-1)$ which contains the site indices of the qubits not to be traced out
[in]	rho	rho is a complex*16 array of dimension $(0{:}2^{**}s-1,0{:}2^{**}s-1)$ which is the input density matrix
[out]	be	be (real*8) is the value of block entropy

Implementation

```
subroutine BERPDMC(s,s1,site,rho,be)
implicit none
integer::s, s1,i,INFOO,site(0:s1-1)
real*8::trace,be,betemp
complex*16::rho(0:2**s-1,0:2**s-1),rdm(0:2**s1-1,0:2**s1-1)
complex*16::WORKK(0:2*(2**s1)-1)
real*8::RWORKK(0:3*(2**s1)-3),WW(0:2**s1-1)
call PTRDMC(s,s1,site,rho,rdm)
call zheev('V','U', 2**s1, rdm, 2**s1, &
                WW, WORKK,2*(2**s1)-1,RWORKK,INFOO)
be=0
betemp=0
do i=0,2**s1-1,1
    if (WW(i)>0.0000000000001) then
        betemp=-(dlog(abs(WW(i)))*abs(WW(i)))/dlog(2.0d0)
        be=be+betemp
    end if
end do
end subroutine
```

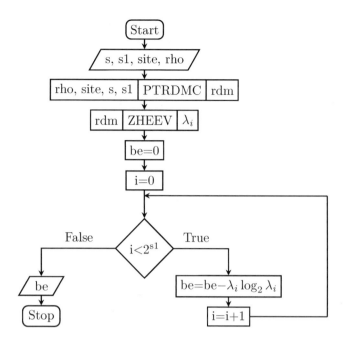

FIGURE 5.15: Flowchart of BERPDMC

Example

In this example we are finding the value of block entropy between the blocks 135 and 246 of the six-qubit density matrix $|\psi_3\rangle\langle\psi_3|$.

```
program example
implicit none
integer,parameter::s=6
integer::i,site(0:2),j
complex*16::psic(0:2**s-1),rhoc(0:2**s-1,0:2**s-1)
real*8::be
do i=0,2**s-1
    psic(i)=dcmplx(i,i+1)
end do
call NORMALIZEVC(psic,2**s)
do i=0,2**s-1
    do j=0,2**s-1
        rhoc(i,j)=psic(i)*conjg(psic(j))
    end do
end do
site(0)=1
site(1)=3
site(2)=5
call BERPDMC(s,3,site,rhoc,be)
write(111,*)be
end program
```

The file "fort.111" contains the required value of block entropy.

8.1419480965576871E-002

5.5   Renyi entropy

Renyi entropy [80,81] S_α is a natural generalization of the Shannon entropy in the discrete case and the von Neumann entropy in the quantum case. It is defined using two quantities, the density matrix and Renyi index $\alpha \geq 0$. Renyi entropy is defined as follows,

$$S_\alpha = \frac{1}{1-\alpha} \log Tr(\rho^\alpha). \tag{5.14}$$

Here α varies from 0 to ∞ such that $\alpha \neq 1$, ρ is a density matrix or reduced density matrix describing the state of the system. In the limit $\alpha \to 1$, The Renyi entropy approaches the von Neumann entanglement entropy. Renyi entropy is used in many areas of quantum information theory.

5.5.1   For a real density matrix

```
subroutine RERPDMR(s,rho,alpha,re)
```

Parameters

In/Out	Argument	Description
[in]	s	s (integer) is the total number of qubits
[in]	rho	rho is a real*8 array of dimension (0:2**s−1,0:2**s−1) which is the input density matrix
[in]	alpha	alpha (real*8) is the value of Renyi index
[out]	re	re (real*8) is the value of Renyi entropy

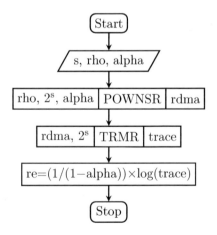

FIGURE 5.16: Flowchart of RERPDMR

Implementation

```
subroutine RERPDMR(s,rho,alpha,re)
implicit none
integer::s
real*8::rho(0:2**s-1,0:2**s-1)
real*8::alpha,re
complex*16::rdma(0:2**s-1,0:2**s-1),trace
call POWNSR(rho,2**s,alpha,rdma)
call TRMC(rdma,2**s,trace)
re=(1/(1-alpha))*dlog(abs(trace))
end subroutine
```

Example

In this example we are finding the value of Renyi entropy of a two-qubit generalized Wener state $\rho(0.5)$.

```
program example
implicit none
integer,parameter::s=2
real*8::re,rho(0:2**s-1,0:2**s-1)
call NQBWERNER(2,0.5d0,rho)
call RERPDMR(s,rho,6.0d0,re)
write(111,*)re
end program
```

The file "fort.111" contains the required value of Renyi entropy.

```
0.56396595878081113
```

5.5.2   For a complex density matrix

```
subroutine RERPDMC(s,s1,site,rho,alpha,re)
```

Parameters

In/Out	Argument	Description
[in]	s	s (integer) is the total number of qubits
[in]	rho	rho is a complex*16 array of dimension $(0:2**s-1,0:2**s-1)$ which is the input density matrix
[in]	alpha	alpha (real*8) is the value of Renyi index
[out]	re	re (real*8) is the value of Renyi entropy

Implementation

```
subroutine RERPDMC(s,rho,alpha,re)
implicit none
integer::s
real*8::re,alpha
complex*16::rho(0:2**s-1,0:2**s-1),trace
complex*16::rdma(0:2**s-1,0:2**s-1)
call POWNHC(rho,2**s,alpha,rdma)
call TRMC(rdma,2**s,trace)
re=(1/(1-alpha))*log(abs(trace))
end subroutine
```

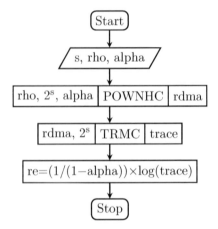

FIGURE 5.17: Flowchart of RERPDMC

Example

In this example we are finding the value of Renyi entropy of the density matrix $\rho_1 \otimes \rho_2$.

```
program example
implicit none
integer,parameter::s=2
complex*16::rho(0:2**s-1,0:2**s-1)
real*8::re
rho(0,0)=dcmplx(0.16d0,0.0d0)
rho(0,1)=dcmplx(-0.04d0,-0.16d0)
rho(0,2)=dcmplx(-0.08d0,-0.04d0)
rho(0,3)=dcmplx(-0.02d0,0.09d0)
rho(1,0)=dcmplx(-0.04d0,0.16d0)
rho(1,1)=dcmplx(0.24d0,0.0d0)
rho(1,2)=dcmplx(0.06d0,-0.07d0)
rho(1,3)=dcmplx(-0.12d0,-0.06d0)
rho(2,0)=dcmplx(-0.08d0,0.04d0)
rho(2,1)=dcmplx(0.06d0,0.07d0)
rho(2,2)=dcmplx(0.24d0,0.0d0)
rho(2,3)=dcmplx(-0.06d0,-0.24d0)
```

```
rho(3,0)=dcmplx(-0.02d0,-0.09d0)
rho(3,1)=dcmplx(-0.12d0,0.06d0)
rho(3,2)=dcmplx(-0.06d0,0.24d0)
rho(3,3)=dcmplx(0.36d0,0.0d0)
call RERPDMC(s,rho,100.0d0,re)
write(111,*)re
end program
```

The file "fort.111" contains the required value of Renyi entropy.

0.37696668017247514

5.6   Negativity and logarithmic negativity

Negativity [82] is a measure of entanglement based on the partial transpose operation. Negativity $\mathcal{N}(\rho)$ for a given bipartite pure or mixed state density matrix ρ is defined with respect to the partial transpose on one of its subsystems A as,

$$\mathcal{N}(\rho) = \frac{\| \rho^{T_A} \| - 1}{2}. \tag{5.15}$$

Negativity is an entanglement monotone and hence a valid measure of entanglement. It measures the degree to which ρ^{T_A} fails to be positive, and therefore it can be regarded as a quantitative version of Peres' criterion for separability (PPT criteria). Associated with negativity, we also define what is called the logarithmic negativity [83] which is $\mathcal{E}_N(\rho) = \log_2(\| \rho^{T_A} \|)$, as expected of a log function, logarithmic negativity has additive properties.

5.6.1   For a real state

```
subroutine NEGVR(s,s1,site,vin,neg,logneg)
```

Parameters

In/Out	Argument	Description
[in]	s	s (integer) is the total number of qubits
[in]	s1	s1 (integer) is the total number of qubits to be transposed
[in]	site	site is an integer array of dimension $(0:s1-1)$ which contains the site indices of the qubits to be transposed
[in]	vin	vin is a real*8 array of dimension $(0:2**s-1)$ which is the input state
[out]	neg	neg (real*8) is the value of negativity
[out]	logneg	logneg (real*8) is the value of logarithmic negativity

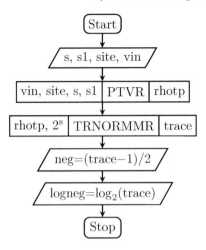

FIGURE 5.18: Flowchart of NEGVR

Implementation

```
subroutine NEGVR(s,s1,site,vin,neg,logneg)
implicit none
integer::s,s1
real*8::trace,neg,logneg
integer,dimension(0:s1-1)::site
real*8,dimension(0:2**s-1)::vin
real*8,dimension(0:2**s-1,0:2**s-1):: rho,rhotp
call PTVR(s,s1,site,vin,rhotp)
call TRNORMMR(rhotp,2**s,trace)
neg=(trace-1)/2.0d0
logneg=dlog(trace)/dlog(2.0d0)
end subroutine
```

Example

In this example we are finding the value of negativity and logarithmic negativity between the blocks 135 and 246 of the six-qubit state $|\psi_2\rangle$.

```
program example
implicit none
integer,parameter::s=6
integer::i,site(0:2)
real*8::psir(0:2**s-1),neg,logneg
do i=0,2**s-1
    psir(i)=i+1
end do
call NORMALIZEVR(psir,2**s)
site(0)=0
site(1)=2
site(2)=4
call NEGVR(6,3,site,psir,neg,logneg)
```

```
write(111,*)neg,logneg
end program
```

The file "fort.111" contains the required values of negativity and logarithmic negativity.

```
9.7674418604651203E-002 0.25743169960441792
```

5.6.2   For a complex state

```
subroutine NEGVC(s,s1,site,vin,neg,logneg)
```

Parameters

In/Out	Argument	Description
[in]	s	s (integer) is the total number of qubits
[in]	s1	s1 (integer) is the total number of qubits to be transposed
[in]	site	site is an integer array of dimension $(0{:}s1{-}1)$ which contains the site indices of the qubits to be transposed
[in]	vin	vin is a complex*16 array of dimension $(0{:}2^{**}s{-}1)$ which is the input state
[out]	neg	neg (real*8) is the value of negativity
[out]	logneg	logneg (real*8) is the value of logarithmic negativity

Implementation

```
subroutine NEGVC(s,s1,site,vin,neg,logneg)
implicit none
integer::s,s1
real*8::trace,neg,logneg
integer,dimension(0:s1-1)::site
complex*16,dimension(0:2**s-1,0:2**s-1)::rho,rhotp
complex*16,dimension(0:2**s-1)::vin
call PTVC(s,s1,site,vin,rhotp)
call TRNORMMC(rhotp,2**s,trace)
neg=(trace-1)/2.0d0
logneg=dlog(trace)/dlog(2.0d0)
end subroutine
```

Example

In this example we are finding the value of negativity and logarithmic negativity between the blocks 135 and 246 of the six-qubit state $|\psi_3\rangle$.

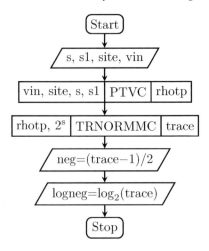

FIGURE 5.19: Flowchart of NEGVC

```
program example
implicit none
integer,parameter::s=6
integer::i,site(0:2)
complex*16::psic(0:2**s-1)
real*8::neg,logneg
do i=0,2**s-1
    psic(i)=dcmplx(i,i+1)
end do
call NORMALIZEVC(psic,2**s)
site(0)=0
site(1)=2
site(2)=4
call NEGVC(s,3,site,psic,neg,logneg)
write(111,*)neg,logneg
end program
```

The file "fort.111" contains the required values of negativity and logarithmic negativity.

```
    9.9963383376052817E-002  0.26294635877747374
```

5.6.3   For a real density matrix

```
subroutine NEGDMR(s,s1,site,rho,neg,logneg)
```

Parameters

In/Out	Argument	Description
[in]	s	s (integer) is the total number of qubits
[in]	s1	s1 (integer) is the total number of qubits to be transposed
[in]	site	site is an integer array of dimension $(0{:}s1-1)$ which contains the site indices of the qubits to be transposed
[in]	rho	rho is a real*8 array of dimension $(0{:}2^{**}s-1,0{:}2^{**}s-1)$ which is the input density matrix
[out]	neg	neg (real*8) is the value of negativity
[out]	logneg	logneg (real*8) is the value of logarithmic negativity

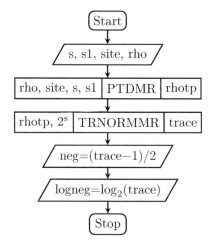

FIGURE 5.20: Flowchart of NEGDMR

Implementation

```
subroutine NEGDMR(s,s1,site,rho,neg,logneg)
implicit none
integer::s,s1
real*8::trace,neg,logneg
integer,dimension(0:s1-1)::site
real*8,dimension(0:2**s-1,0:2**s-1):: rho,rhotp
call PTDMR(s,s1,site,rho,rhotp)
call TRNORMMR(rhotp,2**s,trace)
neg=(trace-1)/2.0d0
logneg=dlog(trace)/dlog(2.0d0)
end subroutine
```

Example

In this example we are finding the value of negativity and logarithmic negativity between the block 135 and 246 of the six-qubit state $|\psi_2\rangle\langle\psi_2|$.

```
program example
implicit none
integer,parameter::s=6
integer::i,site(0:2),j
real*8::psir(0:2**s-1),neg,logneg,rhor(0:2**s-1,0:2**s-1)
do i=0,2**s-1
    psir(i)=i+1
end do
call NORMALIZEVR(psir,2**s)
do i=0,2**s-1
    do j=0,2**s-1
        rhor(i,j)=psir(i)*psir(j)
    end do
end do
site(0)=1
site(1)=3
site(2)=5
call NEGDMR(s,3,site,rhor,neg,logneg)
write(111,*)neg,logneg
end program
```

The file "fort.111" contains the required values of negativity and logarithmic negativity.

9.7674418604651203E-002 0.25743169960441792

5.6.4 For a complex density matrix

```
subroutine NEGDMC(s,s1,site,rho,neg,logneg)
```

Parameters

In/Out	Argument	Description
[in]	s	s (integer) is the total number of qubits
[in]	s1	s1 (integer) is the total number of qubits to be transposed
[in]	site	site is an integer array of dimension $(0:s1-1)$ which contains the site indices of the qubits to be transposed
[in]	rho	rho is a complex*16 array of dimension $(0:2**s-1,0:2**s-1)$ which is the input density matrix
[out]	neg	neg (real*8) is the value of negativity
[out]	logneg	logneg (real*8) is the value of logarithmic negativity

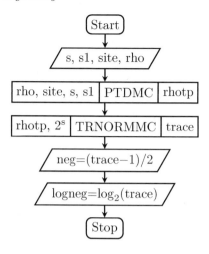

FIGURE 5.21: Flowchart of NEGDMC

Implementation

```
subroutine NEGDMC(s,s1,site,rho,neg,logneg)
implicit none
integer::s,s1
real*8::trace,neg,logneg
integer,dimension(0:s1-1)::site
complex*16,dimension(0:2**s-1,0:2**s-1):: rho,rhotp
call PTDMC(s,s1,site,rho,rhotp)
call TRNORMMC(rhotp,2**s,trace)
neg=(trace-1)/2.0d0
logneg=dlog(trace)/dlog(2.0d0)
end subroutine
```

Example

In this example we are finding the value of negativity and logarithmic negativity between the blocks 135 and 246 of the six-qubit state $|\psi_3\rangle\langle\psi_3|$.

```
program example
implicit none
integer,parameter::s=6
integer::i,site(0:2),j
complex*16::psic(0:2**s-1),rhoc(0:2**s-1,0:2**s-1)
real*8::neg,logneg
do i=0,2**s-1
    psic(i)=dcmplx(i,i+1)
end do
call NORMALIZEVC(psic,2**s)
do i=0,2**s-1
    do j=0,2**s-1
        rhoc(i,j)=psic(i)*conjg(psic(j))
    end do
```

```
end do
site(0)=1
site(1)=3
site(2)=5
call NEGDMC(s,3,site,rhoc,neg,logneg)
write(111,*)neg,logneg
end program
```

The file "fort.111" contains the required values of negativity and logarithmic negativity.

```
9.9963383376052817E-002 0.26294635877747374
```

5.7 Q measure or the Meyer-Wallach-Brennen measure

Q measure [84, 85] is an entanglement measure for a multipartite pure state in the Hilbert space of dimension $(\mathbb{C}^2)^{\otimes N}$, it is defined as,

$$Q = 2\left(1 - \frac{1}{N}\sum_{i=1}^{N} Tr(\rho_i^2)\right). \tag{5.16}$$

Where ρ_i is a single qubit density matrix got by tracing all the other qubits except the ith qubit and N is the total number of qubits. From the above expression it is clear that, the Q measure is the average subsystem linear entropy of the constituent qubits. The bound on Q is such that, $0 \le Q \le 1$.

5.7.1 For a real state

```
subroutine QMVR(s,vin,Q)
```

Parameters

In/Out	Argument	Description
[in]	s	s (integer) is the total number of qubits
[in]	vin	vin is a real*8 array of dimension (0:2**s−1) which is the input state
[out]	Q	Q (real*8) is the value of Q measure

Implementation

```
subroutine QMVR(s,vin,Q)
implicit none
integer::s,site(0:0),i
real*8::vin(0:2**s-1),rdm(0:1,0:1)
```

```
real*8::rdm2(0:1,0:1),trace,summ,Q
summ=0.0d0
do i=0,s-1,1
    site(0)=i
    call PTRVR(s,1,site,vin,rdm)
    rdm2=matmul(rdm,rdm)
    call TRMR(rdm2,2,trace)
    summ=summ+trace
end do
Q=2*(1-summ/s)
end subroutine
```

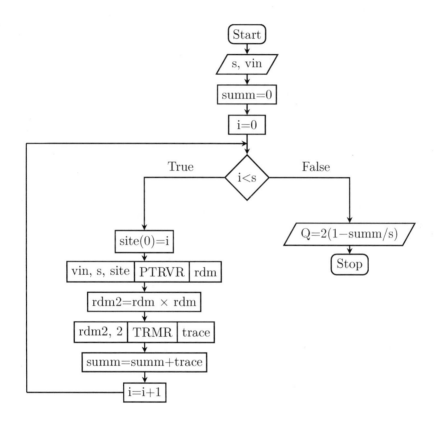

FIGURE 5.22: Flowchart of QMVR

Example

In this example we are finding the value of Q measure of the six-qubit state $|\psi_2\rangle$.

```
program example
implicit none
integer,parameter::s=6
integer::i
real*8::psir(0:2**s-1),Q
do i=0,2**s-1
```

```
   psir(i)=i+1
end do
call NORMALIZEVR(psir,2**s)
call QMVR(s,psir,Q)
write(111,*)Q
end program
```

The file "fort.111" contains the required value of Q measure.

```
1.5888838041352216E-002
```

5.7.2   For a complex state

```
subroutine QMVC(s,vin,Q)
```

Parameters

In/Out	Argument	Description
[in]	s	s (integer) is the total number of qubits
[in]	vin	vin is a complex*16 array of dimension (0:2**s−1) which is the input state
[out]	Q	Q (real*8) is the value of Q measure

Implementation

```
subroutine QMVC(s,vin,Q)
implicit none
integer::s,site(0:0),i
complex*16::vin(0:2**s-1),trace
complex*16,dimension(0:1,0:1)::rdm,rdm2
real*8::summ,Q
summ=0.0d0
do i=0,s-1,1
    site(0)=i
    call PTRVC(s,1,site,vin,rdm)
    rdm2=matmul(rdm,rdm)
    call TRMC(rdm2,2,trace)
    summ=summ+real(trace)
end do
Q=2*(1-summ/s)
end subroutine
```

Example

In this example we are finding the value of Q measure of the six-qubit state $|\psi_3\rangle$.

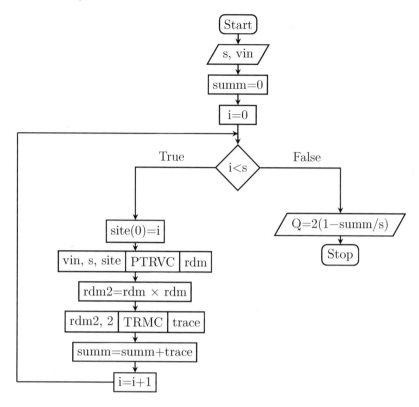

FIGURE 5.23: Flowchart of QMVC

```
program example
implicit none
integer,parameter::s=6
integer::i
complex*16::psic(0:2**s-1)
real*8::Q
do i=0,2**s-1
    psic(i)=dcmplx(i,i+1)
end do
call NORMALIZEVC(psic,2**s)
call QMVC(s,psic,Q)
write(111,*)Q
end program
```

The file "fort.111" contains the required value of Q measure.

```
1.6642262287920317E-002
```

5.7.3   For a real density matrix

```
subroutine QMDMR(s,rho,Q)
```

Parameters

In/Out	Argument	Description
[in]	s	s (integer) is the total number of qubits
[in]	rho	rho is a real*8 array of dimension $(0{:}2^{**}s{-}1,0{:}2^{**}s{-}1)$ which is the input pure state density matrix
[out]	Q	Q (real*8) is the value of Q measure

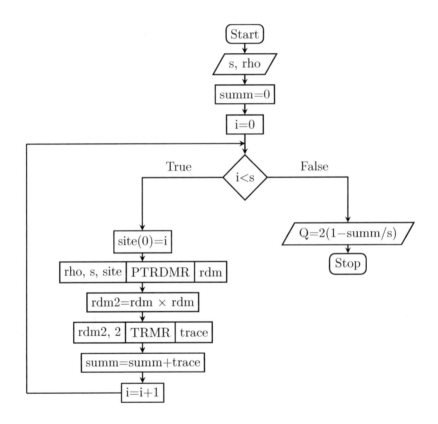

FIGURE 5.24: Flowchart of QMDMR

Implementation

```
subroutine QMDMR(s,rho,Q)
implicit none
integer::s,site(0:0),i
real*8::rho(0:2**s-1,0:2**s-1),rdm(0:1,0:1)
real*8::rdm2(0:1,0:1),trace,summ,Q
summ=0.0d0
do i=0,s-1,1
    site(0)=i
    call PTRDMR(s,1,site,rho,rdm)
```

```
    rdm2=matmul(rdm,rdm)
    call TRMR(rdm2,2,trace)
    summ=summ+trace
end do
Q=2*(1-summ/s)
end subroutine
```

Example

In this example we are finding the value of Q measure of the six-qubit density matrix $|\psi_2\rangle\langle\psi_2|$.

```
program example
implicit none
integer,parameter::s=6
integer::i,j
real*8::psir(0:2**s-1),Q,rhor(0:2**s-1,0:2**s-1)
do i=0,2**s-1
    psir(i)=i+1
end do
call NORMALIZEVR(psir,2**s)
do i=0,2**s-1
    do j=0,2**s-1
        rhor(i,j)=psir(i)*psir(j)
    end do
end do
call QMDMR(s,rhor,Q)
write(111,*)Q
end program
```

The file "fort.111" contains the required value of Q measure.

```
  1.5888838041352216E-002
```

5.7.4 For a complex density matrix

```
subroutine QMDMC(s,rho,Q)
```

Parameters

In/Out	Argument	Description
[in]	s	s (integer) is the total number of qubits
[in]	rho	rho is a complex*16 array of dimension (0:2**s−1,0:2**s−1) which is the input pure state density matrix
[out]	Q	Q (real*8) is the value of Q measure

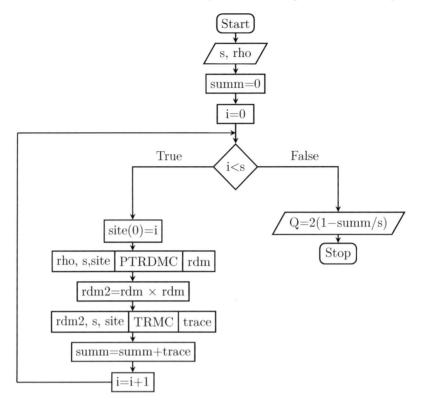

FIGURE 5.25: Flowchart of QMDMC

Implementation

```
subroutine QMDMC(s,rho,Q)
implicit none
integer::s,site(0:0),i
complex*16::rho(0:2**s-1,0:2**s-1),trace
complex*16,dimension(0:1,0:1)::rdm,rdm2
real*8::summ,Q
summ=0.0d0
do i=0,s-1,1
    site(0)=i
    call PTRDMC(s,1,site,rho,rdm)
    rdm2=matmul(rdm,rdm)
    call TRMC(rdm2,2,trace)
    summ=summ+real(trace)
end do
Q=2*(1-summ/s)
end subroutine
```

Example

In this example we are finding the value of Q measure of the six-qubit density matrix $|\psi_3\rangle\langle\psi_3|$.

```
program example
implicit none
integer,parameter::s=6
integer::i,j
complex*16::psic(0:2**s-1),rhoc(0:2**S-1,0:2**s-1)
real*8::Q
do i=0,2**s-1
    psic(i)=dcmplx(i,i+1)
end do
call NORMALIZEVC(psic,2**s)
do i=0,2**s-1
    do j=0,2**s-1
        rhoc(i,j)=psic(i)*conjg(psic(j))
    end do
end do
call QMDMC(s,rhoc,Q)
write(111,*)Q
end program
```

The file "fort.111" contains the required value of Q measure.

```
1.6642262287920317E-002
```

5.8 Entanglement spectrum

The entanglement spectrum [86,87] is the negative natural logarithm of the eigenvalue spectrum of the reduced density matrix. Instead of having a number (von Neumann entropy) for quantifying entanglement, Haldane in his seminal work pointed out that the spectrum of the reduced density matrix contains more information regarding the entanglement. This quantity is more important when we study condensed matter systems like topological materials, quantum hall effect, etc. If we have a reduced density matrix of a composite system AB as ρ_A, then the entanglement spectrum is defined as $-ln(\lambda_i)$, where λ_i with $i = 1, \cdots, \dim(\rho_A)$ are the eigenvalues of ρ_A.

5.8.1 For a real density matrix

```
subroutine ENTSPECDMR(s,rho,eval,lneval)
```

Parameters

In/Out	Argument	Description
[in]	s	s (integer) is the total number of qubits
[in]	rho	rho is a real*8 array of dimension $(0{:}2^{**}s{-}1,0{:}2^{**}s{-}1)$ which is the input density matrix
[out]	eval	eval is a real*8 array of dimension $(0{:}2^{**}s{-}1)$ which are the eigenvalues of rho
[out]	lneval	lneval is a real*8 array of dimension $(0{:}2^{**}s{-}1)$ which is the negative logarithm of the eigenvalues of rho

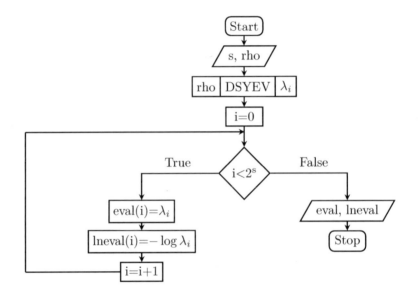

FIGURE 5.26: Flowchart of ENTSPECDMR

Implementation

```
subroutine ENTSPECDMR(s,rho,eval,lneval)
integer::s,i,INFO
real*8::rho(0:2**s-1,0:2**s-1),eval(0:2**s-1)
real*8::WORK(0:3*(2**s)-1),W(0:2**s-1),lneval(0:2**s-1)
call DSYEV('N','U', 2**s, rho, 2**s, W, WORK,3*(2**s)-1,INFO)
do i=0,2**s-1
    eval(i)=W(i)
    lneval(i)=-log(W(i))
end do
end subroutine
```

Example

In this example we are finding entanglement spectrum of a two-qubit generalized Wener state $\rho(0.5)$.

```fortran
program example
real*8::rho(0:3,0:3),eval(0:3),lneval(0:3)
integer::i
call NQBWERNER(2,0.5d0,rho)
call ENTSPECDMR(2,rho,eval,lneval)
do i=0,3
   write(111,*)eval(i),lneval(i)
end do
end program
```

The file "fort.111" contains the eigenvalues and the entanglement spectrum in the column one and two, respectively.

0.12499999999999999	2.0794415416798362
0.12500000000000000	2.0794415416798357
0.12500000000000000	2.0794415416798357
0.62499999999999989	0.47000362924573574

5.8.2 For a complex density matrix

```fortran
subroutine ENTSPECDMC(s,rho,eval,lneval)
```

Parameters

In/Out	Argument	Description
[in]	s	s (integer) is the total number of qubits
[in]	rho	rho is a complex*16 array of dimension $(0:2**s-1,0:2**s-1)$ which is the input density matrix
[out]	eval	eval is a real*8 array of dimension $(0:2**s-1)$ which are the eigenvalues of rho
[out]	lneval	lneval is a real*8 array of dimension $(0:2**s-1)$ which is the negative logarithm of the eigenvalues of rho

Implementation

```fortran
subroutine ENTSPECDMC(s,rho,eval,lneval)
integer::s,i,INFO
```

```
complex*16::rho(0:2**s-1,0:2**s-1)
real*8::eval(0:2**s-1),lneval(0:2**s-1)
complex*16::WORK(0:2*(2**s)-1)
real*8::RWORK(0:3*(2**s)-3),W(0:2**s-1)
call  zheev('N','U', 2**s, rho, 2**s, W, WORK, &
                  2*(2**s)-1,RWORK,INFO)
do i=0,2**s-1
    eval(i)=W(i)
    lneval(i)=-log(W(i))
end do
end subroutine
```

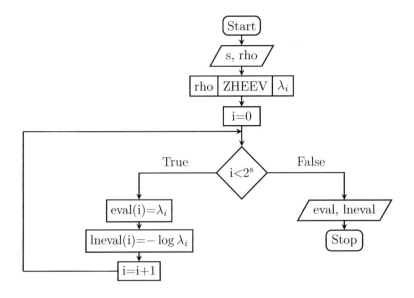

FIGURE 5.27: Flowchart of ENTSPECDMC

Example

In this example we are finding entanglement spectrum of the density matrix $\rho_1 \otimes \rho_2$.

```
program example
real*8::eval(0:3),lneval(0:3)
complex*16::rho(0:3,0:3)
integer::i
rho(0,0)=dcmplx(0.16d0,0.0d0)
rho(0,1)=dcmplx(-0.04d0,-0.16d0)
rho(0,2)=dcmplx(-0.08d0,-0.04d0)
rho(0,3)=dcmplx(-0.02d0,0.09d0)
rho(1,0)=dcmplx(-0.04d0,0.16d0)
rho(1,1)=dcmplx(0.24d0,0.0d0)
rho(1,2)=dcmplx(0.06d0,-0.07d0)
rho(1,3)=dcmplx(-0.12d0,-0.06d0)
rho(2,0)=dcmplx(-0.08d0,0.04d0)
rho(2,1)=dcmplx(0.06d0,0.07d0)
```

```
rho(2,2)=dcmplx(0.24d0,0.0d0)
rho(2,3)=dcmplx(-0.06d0,-0.24d0)
rho(3,0)=dcmplx(-0.02d0,-0.09d0)
rho(3,1)=dcmplx(-0.12d0,0.06d0)
rho(3,2)=dcmplx(-0.06d0,0.24d0)
rho(3,3)=dcmplx(0.36d0,0.0d0)
call ENTSPECDMC(2,rho,eval,lneval)
do i=0,3
    write(111,*)eval(i),lneval(i)
end do
end program
```

The file "fort.111" contains the eigenvalues and the entanglement spectrum in the column one and two, respectively.

1.9316526959009470E-002	3.9467942303836794
5.6419404329061992E-002	2.8749421312615748
0.23573449876267280	1.4450491124928539
0.68852956994925574	0.37319701337075006

5.9 Residual entanglement for three qubits

Residual entanglement of a three qubit pure state is also known as three tangle [88]. We define the three tangle to be,

$$\tau_3 = \tau_{1(23)} - \tau_{12} - \tau_{13}. \tag{5.17}$$

Where τ can be defined as the square of concurrence. We can also define the tangle between one qubit (say the ith) and the rest of the qubits if the total state is pure. This is because, of Schmidt decomposition. The tangle between qubit i and the rest is the one tangle given by $\tau_{i,(\text{rest of the spins})} = 4\det(\rho_i)$, where ρ_i is the single qubit reduced density matrix got by partial tracing all other qubits except the ith qubit. This concept can be used to measure the entanglement in a three qubit system like the famous GHZ state or the W state. We can rewrite Eq. [5.17] as follows,

$$\tau_3 = 4\det(\rho_1) - C_{12}^2 - C_{13}^2. \tag{5.18}$$

From the structure of the GHZ and W state we see the contrasting behavior of bipartite entanglement in them, this three tangle will be useful in differentiating these entanglement properties for such three qubit pure states.

5.9.1 For a real state

```
subroutine RESENTVR(vin,t)
```

Parameters

In/Out	Argument	Description
[in]	vin	vin is a real*8 array of dimension (0:7) which is the input pure state
[out]	t	t (real*8) is the value of the residual entanglement

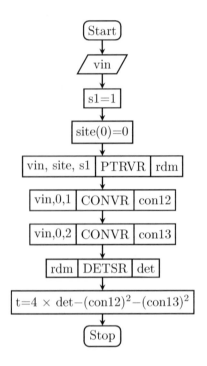

FIGURE 5.28: Flowchart of RESENTVR

Implementation

```
subroutine RESENTVR(vin,t)
integer,parameter::s=3,s1=1
real*8::vin(0:2**s-1),rdm(0:1,0:1),det,t,con12,con13
integer::site(0:s1-1)
site(0)=0
call PTRVR(s,s1,site,vin,rdm)
call CONVR(s,0,1,vin,con12)
call CONVR(s,0,2,vin,con13)
call DETSR(rdm,2,det)
t=4*det-con12**2-con13**2
end subroutine
```

Example

In this example we are finding the value of the residual entanglement of the three-qubit state $|\psi_2\rangle$.

```
program example
integer,parameter::s=3
real*8::psi(0:2**s-1),t
do i=0,2**s-1
    psi(i)=i+1
end do
call NORMALIZEVR(psi,2**s)
call RESENTVR(psi,t)
write(111,*)t
end program
```

The file "fort.111" contains the required value of the residual entanglement.

```
-3.9031278209478160E-017
```

5.9.2   For a complex vector

```
subroutine RESENTVC(vin,t)
```

Parameters

In/Out	Argument	Description
[in]	vin	vin is a complex*16 array of dimension (0:7) which is the input pure state
[out]	t	t (real*8) is the value of the residual entanglement

Implementation

```
subroutine RESENTVC(vin,t)
integer,parameter::s=3,s1=1
complex*16::vin(0:2**s-1),rdm(0:1,0:1)
real*8::t,con12,con13,det
integer::site(0:s1-1)
site(0)=0
call PTRVC(s,s1,site,vin,rdm)
call CONVC(s,0,1,vin,con12)
call CONVC(s,0,2,vin,con13)
call DETHC(rdm,2,det)
t=4*det-con12**2-con13**2
end subroutine
```

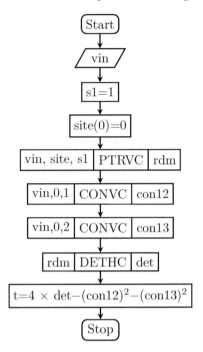

FIGURE 5.29: Flowchart of RESENTVC

Example

In this example we are finding the value of the residual entanglement of the three-qubit state $|\psi_3\rangle$.

```
program example
implicit none
integer,parameter::s=3
integer::i
complex*16::psic(0:2**s-1)
real*8::t
do i=0,2**s-1
    psic(i)=dcmplx(i,i+1)
end do
call NORMALIZEVC(psic,2**s)
call RESENTVC(psic,t)
write(111,*)t
end program
```

The file "fort.111" contains the required value of the residual entanglement.

```
1.3877787807814457E-017
```

5.9.3 For a real density matrix

```
subroutine RESENTDMR(rho,t)
```

Parameters

In/Out	Argument	Description
[in]	rho	rho is a real*8 array of dimension (0:7,0:7) which is the input pure state density matrix
[out]	t	t (real*8) is the value of the residual entanglement

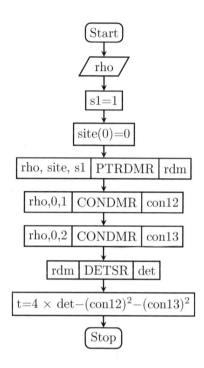

FIGURE 5.30: Flowchart of RESENTDMR

Implementation

```
subroutine RESENTDMR(rho,t)
integer,parameter::s=3,s1=1
real*8::rho(0:2**s-1,0:2**s-1),rdm(0:1,0:1),det,t,con12,con13
integer::site(0:s1-1)
site(0)=0
call PTRDMR(s,s1,site,rho,rdm)
call CONDMR(s,0,1,rho,con12)
call CONDMR(s,0,2,rho,con13)
call DETSR(rdm,2,det)
t=4*det-con12**2-con13**2
end subroutine
```

Example

In this example we are finding the value of the residual entanglement of the three-qubit density matrix $|\psi_2\rangle\langle\psi_2|$.

```fortran
program example
implicit none
integer,parameter::s=3
integer::i,j
real*8::psir(0:2**s-1),t,rhor(0:2**s-1,0:2**s-1)
do i=0,2**s-1
    psir(i)=i+1
end do
call NORMALIZEVR(psir,2**s)
do i=0,2**s-1
    do j=0,2**s-1
        rhor(i,j)=psir(i)*psir(j)
    end do
end do
call RESENTDMR(rhor,t)
write(111,*)t
end program
```

The file "fort.111" contains the required value of the residual entanglement.

```
-3.9031278209478160E-017
```

5.9.4   For a complex density matrix

```fortran
subroutine RESENTDMC(rho,t)
```

Parameters

In/Out	Argument	Description
[in]	rho	rho is a complex*16 array of dimension (0:7,0:7) which is the input pure state density matrix
[out]	t	t (real*8) is the value of the residual entanglement

Implementation

```fortran
subroutine RESENTDMC(rho,t)
integer,parameter::s=3,s1=1
complex*16::rho(0:2**s-1,0:2**s-1),rdm(0:1,0:1)
real*8::t,con12,con13,det
integer::site(0:s1-1)
site(0)=0
call PTRDMC(s,s1,site,rho,rdm)
call CONDMC(s,0,1,rho,con12)
```

```
call CONDMC(s,0,2,rho,con13)
call DETHC(rdm,2,det)
t=4*det-con12**2-con13**2
end subroutine
```

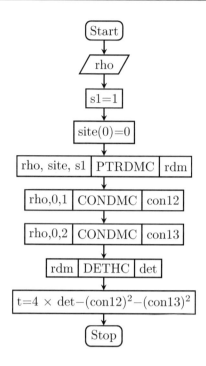

FIGURE 5.31: Flowchart of RESENTDMC

Example

In this example we are finding the value of the residual entanglement of the three-qubit density matrix $|\psi_3\rangle\langle\psi_3|$.

```
program example
implicit none
integer,parameter::s=3
integer::i,j
complex*16::psic(0:2**s-1),rhoc(0:2**S-1,0:2**s-1)
real*8::t
do i=0,2**s-1
    psic(i)=dcmplx(i,i+1)
end do
call NORMALIZEVC(psic,2**s)
do i=0,2**s-1
    do j=0,2**s-1
        rhoc(i,j)=psic(i)*conjg(psic(j))
    end do
end do
call RESENTDMC(rhoc,t)
write(111,*)t
```

```
end program
```

The file "fort.111" contains the required value of the residual entanglement.

```
1.3877787807814457E-017
```

5.10   Detection of entanglement

5.10.1   PPT criteria or the Peres positive partial transpose criteria

For any bipartite separable state ρ_{AB}, if we perform a partial transpose on one of the subsytems, say B, to obtain ρ^{T_B}, then we say ρ is separable if all the eigenvalues of ρ^{T_B} are non-negative. If ρ^{T_B} has even one negative eigenvalue, then ρ_{AB} is entangled. In the 2×2 and 2×3 dimensions the condition is both necessary and also sufficient [69, 89]. This method of entanglement detection is very important, as for a mixed state, the absence of Schmidt decomposition coupled with the non-uniqueness in the pure state decomposition of the mixed state makes PPT a noteworthy criteria. In the recipes below, we give the separability criteria for only 2×2 systems. We have given the example of PPT criteria using partial transpose routines as follows. In what follows, we will detect entanglement in the two-qubit generalized Werner state $\rho(0.5)$. As we already know that, for $p > 1/3$ the Werner state is non-separable state.

```
program example
implicit none
integer,parameter::s=2,s1=1
integer::site(0:s1-1),i,INFO
real*8::rho(0:2**s-1,0:2**s-1),rhotp(0:2**s-1,0:2**s-1)
real*8::W(0:2**s-1)
real*8::WORK(0:3*(2**s)-2)
character*3::test
test='PPT'
call NQBWERNER(2,0.5d0,rho)
call PTDMR(s,s1,site,rho,rhotp)
call DSYEV('N','U', 2**s, rhotp, 2**s, &
                W, WORK,3*(2**s)-1,INFO)
do i=0,2**s-1,1
   if(W(i) .lt. 0.00000000) then
       test='NPT'
       exit
   end if
end do
write(111,*)test
end program
```

The file "fort.111" will contain NPT, meaning it is not separable.

5.10.2 Reduction criteria

For any seperable bipartite state ρ_{AB} such that $\rho_{AB} \in \mathcal{H}_{N_A \times N_B}$, where N_A and N_B are the dimensions of the subsystem spaces A and B, respectively. We then say that, ρ_{AB} is separable only if the following operators are positive, meaning their eigenvalues are non-negative.

$$\rho^A \otimes \mathbb{I} - \rho_{AB} > 0, \tag{5.19}$$

$$\mathbb{I} \otimes \rho^B - \rho_{AB} > 0. \tag{5.20}$$

This criteria [90] is a set of two inequalites as shown above where ρ_A and ρ_B are the reduced density matrices of the subsystems A and B, respectively. In the recipes below, we give the separability criteria for only 2×2 systems. In this example we will detect entanglement in two-qubit generalized Werner state $\rho(0.5)$.

```fortran
program example
implicit none
integer,parameter::s=2,s1=1
integer::s2,site(0:s1-1),i,INFO
real*8::rho(0:2**s-1,0:2**s-1),rdm(0:2**s1-1,0:2**s1-1)
real*8::II(0:2**(s-s1)-1,0:2**(s-s1)-1)
real*8::C(0:2**s-1,0:2**s-1),temp(0:2**s-1,0:2**s-1)
real*8::W(0:2**s-1)
real*8::WORK(0:3*(2**s)-2)
character*15::test
site(0)=0
s2=s-s1
test='unentangled'
temp=0.0d0
II=0.0d0
do i=0,2**s2-1
    II(i,i)=1.0d0
end do
call NQBWERNER(2,0.5d0,rho)
call PTRDMR(s,s1,site,rho,rdm)
call KPMR(2**s1,2**s1,rdm,2**s2,2**s2,II,C)
temp=C-rho
call DSYEV('N','U', 2**s, temp, 2**s, W, &
                    WORK,3*(2**s)-1,INFO)
do i=0,2**s-1,1
    if(W(i) .lt. 0.00000000) then
        test='entangled'
        exit
    end if
end do
write(111,*)test
end program
```

The file "fort.111" will contain 'entangled' meaning it is non-separable.

6

One-Dimensional Spin Chain Models in Condensed Matter Theory

In this chapter we will give the numerical recipes for constructing the spin half Fermionic Hamiltonians in one dimension [91–94], basically a lattice of contiguous electrons along a line. Each electron at a given point on the one-dimensional lattice represents one qubit (here qubits are synonymous with spins) of information. Spin Hamiltonians or what is popularly called the spin chain models are prototypical in understanding the origin of magnetism and magnetic phases [95–97]. Not only that, these spin chains at a temperature of zero kelvin can be used to study and understand quantum phase transitions [66,67], which, unlike thermal phase transitions do not involve any thermal parameter. These Hamiltonians also prove as a fertile ground for testing various quantum information tasks not only theoretically but also experimentally [98–100]. We have given the numerical recipes for all possible interaction Hamiltonians and if any spin chain needs to be studied, different recipes can be merged very easily together and we can construct the required Hamiltonian part by part. As we progress through the chapter, we will give some examples on how to construct some famous spin chains using our numerical recipes for the sake of illustration.

To understand further, we consider the simple but the well known Ising model in the external magnetic field along the Z direction [101–103]. The Hamiltonian of this model with N spins (qubits) is given by,

$$H_{ising} = -J \sum_{i=1}^{N} \sigma_i^z \sigma_{i+1}^z - B \sum_{i=1}^{N} \sigma_i^z. \tag{6.1}$$

Where J is the coupling constant and B is the strength of the magnetic field. Every spin is represented by a quantum operator acting upon the tensor product $(\mathbb{C}^2)^{\otimes N}$ as each site is associated with a complex two-dimensional Hilbert space. To make things more clear, let us take a 3 spin Ising model, we can have two different types of boundary conditions, one is the periodic boundary condition (PBC) in which there is a wrapping up of the spins end to end and the other which is the open boundary condition (OBC) where there is no wrapping up and end spins are left free. Let us write the Hamiltonian for the model given in Eq. [6.1] as below, considering periodic boundary conditions with end to end wrap up of the form $\sigma_{N+1} \equiv \sigma_1$

$$
\begin{aligned}
H_{ising} &= -J \sum_{i=1}^{3} \sigma_i^z \sigma_{i+1}^z - B \sum_{i=1}^{3} \sigma_i^z, \tag{6.2} \\
&= -J(\sigma_1^z \sigma_2^z + \sigma_2^z \sigma_3^z + \underbrace{\sigma_3^z \sigma_1^z}_{wrap\ up}) - B(\sigma_1^z + \sigma_2^z + \sigma_3^z). \tag{6.3}
\end{aligned}
$$

Now for a model with open boundary conditions where there is no such condition of end to end wrap up like $\sigma_{N+1} \equiv \sigma_1$, we can write

$$H_{ising} = -J\sum_{i=1}^{3}\sigma_i^z\sigma_{i+1}^z - B\sum_{i=1}^{3}\sigma_i^z, \tag{6.4}$$

$$= -J(\sigma_1^z\sigma_2^z + \sigma_2^z\sigma_3^z) - B(\sigma_1^z + \sigma_2^z + \sigma_3^z). \tag{6.5}$$

However, these are the PBCs and OBCs for nearest-neighbor interaction, if we have r nearest neighbors interacting, there will be r such end to end wrap up terms of the form $\sigma_{N+r} \equiv \sigma_r$ for $r = 1, 2, \ldots$. It is also important to note that $\sigma_i^z = \mathbb{I}^{\otimes i-1} \otimes \sigma^z \otimes \mathbb{I}^{\otimes N-i}$, where $\mathbb{I}$ is the 2×2 identity matrix. This is so because, the total Hamiltonian for N spins will be a $2^N \times 2^N$ matrix in the $(\mathbb{C}^2)^{\otimes N}$- dimensional Hilbert space. All this implies that, the Pauli operator z with a subscript i which is σ_i^z will only act on the qubit at site i. At each site i there is a quantum mechanical spin half particle either pointing along "up $(|\uparrow\rangle)$" or "down $(|\downarrow\rangle)$" which are respectively denoted by $|0\rangle$ and $|1\rangle$. Having the above discussed facts in mind, we can proceed further. All the recipes given below are for both PBC's and OBC's which can be chosen as a user defined variable while execution. The user also can choose till how many neighbours he needs to build the spin chain. Also note that, in this chapter we are using σ matrices instead of the naive spin matrix S, bearing in mind the relationship between them as $S_i = \frac{1}{2}\sigma_i$ (with $\hbar = 1$), with $i = x, y, z$.

Note that for all Hamiltonians from section [6.1] to section [6.3], the output of an example contains only the integer part of the matrix elements, this is done to accommodate the output in the given space. However for the section [6.4] which is the DM interaction, only integer part of the imaginary elements are displayed for the same reason. Note that, these subroutines when used in a program gives the matrix elements as double precision numbers. In this chapter, all examples are given for three spins. In the code given in this chapter we have represented the interaction parameter like magnetic field, interaction strength (direct and anti symmetric) by an array of dimension equal to number of spins in the system.

In the codes given for the interaction of the spins with magnetic field, we have considered B to be an array of dimension equal to the number of spins in the system. This also gives us the freedom to choose an inhomogenenous magnetic field at different sites, meaning, the ith entry of this array will be the magnitude of the applied magnetic field at the ith site. If the magnetic field is homogeneous, all entries of the array B will be same. Similarly, for spin-spin interactions, we have chosen the interaction parameter to be an array of dimension equal to the number of spins in the system. The ith entry of this array will be the interaction strength between the ith and (i+r)th spin. If the interaction is isotropic, all the entries of the corresponding interaction parameter will be same. However, if random interactions are present between the spins, the array will be filled with different numbers (drawn from any probability distribution) corresponding to each pair of spins. You can understand more about usage as you see the examples given as we go through the chapter.

6.1 Hamiltonian of spins interacting with an external magnetic field

The interaction Hamiltonian of a linear array of spin half electrons interacting with an external magnetic field of strength B in general is given by,

$$H = \sum_{i=1}^{N} B_i\sigma_i^j, \tag{6.6}$$

where N is the total number of qubits present in the system and σ^j are the Pauli spin matrices, depending on which direction, the magnetic field is oriented, j can be accordingly labeled as x, y, z as follows. Note that B_i is inside summation to accommodate a general inhomogeneous magnetic field.

6.1.1 Hamiltonian of spins interacting with an external magnetic field in the X direction

In this case we choose $j = x$ in Eq. [6.6] and thereby, the interaction Hamiltonian of a linear array of spins interacting with the external magnetic field in the X direction is given by,

$$H = \sum_{i=1}^{N} B_i \sigma_i^x, \tag{6.7}$$

subroutine XHAM(s,B,HB)

Parameters

In/Out	Argument	Description
[in]	s	s (integer) is the total number of spins
[in]	B	B is a real*8 array of dimension (0:s−1). The ith entry of this array is the strength of the applied magnetic field along X direction on the ith spin
[out]	HB	HB is a real*8 array of dimension (0:2**s−1,0:2**s−1) which is the required Hamiltonian

Implementation

```
subroutine XHAM(s,B,HB)
implicit none
integer::s,t(0:s-1),y(0:s-1),i,j,d,w
real*8::B(0:s-1),h,e
real*8::HB(0:2**s-1,0:2**s-1)
HB=0.0d0
do i=0,2**s-1,1
    do j=i,2**s-1,1
        call DTOB(j,y,s)
        e=0
        t=y
        do d=0,s-1
            t(d)=1-t(d)
            call BTOD(t,w,s)
            if(w==i) then
                h=B(d)
            else
                h=0
```

```
            end if
            e=e+h
            t=y
         end do
         HB(i,j)=e
         HB(j,i)=HB(i,j)
      end do
   end do
end subroutine
```

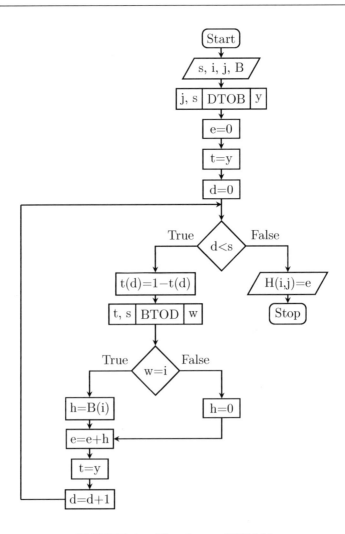

FIGURE 6.1: Flowchart of XHAM

Example

Here, we are finding the Hamiltonian of spins interacting with inhomogeneous magnetic field in x direction.

```
program example
implicit none
```

```
integer,parameter::s=3
integer::i
real*8::H(0:2**s-1,0:2**s-1),B(0:s-1)
B=(/1.0d0,2.0d0,3.0d0/)
call XHAM(s,B,H)
do i=0,2**s-1,1
write(111,*)int(H(i,:))
end do
end program
```

The file "fort.111" contains the matrix elements of the required Hamiltonian.

0	3	2	0	1	0	0	0
3	0	0	2	0	1	0	0
2	0	0	3	0	0	1	0
0	2	3	0	0	0	0	1
1	0	0	0	0	3	2	0
0	1	0	0	3	0	0	2
0	0	1	0	2	0	0	3
0	0	0	1	0	2	3	0

6.1.2 Hamiltonian of spins interacting with an external magnetic field in the Y direction

In this case we choose $j = y$ in Eq. [6.6] and thereby, the interaction Hamiltonian of a linear array of spins interacting with the external magnetic field in the y direction is given by,

$$H = \sum_{i=1}^{N} B_i \sigma_i^y. \tag{6.8}$$

```
subroutine YHAM(s,B,HB)
```

Parameters

In/Out	Argument	Description
[in]	s	s (integer) is the total number of spins
[in]	B	B is a real*8 array of dimension (0:s−1). The ith entry of this array is the strength of the applied magnetic field along Y direction on the ith spin
[out]	HB	HB is a complex*16 array of dimension (0:2**s−1,0:2**s−1) which is the required Hamiltonian

Implementation

```
subroutine YHAM(s,B,HB)
implicit none
```

```fortran
integer::s,t(0:s-1),y(0:s-1),i,j,d,w
real*8::B(0:s-1)
complex*16::HB(0:2**s-1,0:2**s-1),phase,h,e
HB=0.0d0
do i=0,2**s-1,1
    do j=i,2**s-1,1
        call DTOB(j,y,s)
        e=0
        t=y
        do d=0,s-1
            if(t(d)==0) then
                t(d)=1
                phase=dcmplx(0,1)
            else
                t(d)=0
                phase=dcmplx(0,-1)
            end if
            call BTOD(t,w,s)
            if(w==i) then
                h=phase*B(d)
            else
                h=0
            end if
            e=e+h
            t=y
        end do
        HB(i,j)=e
        HB(j,i)=conjg(HB(i,j))
    end do
end do
end subroutine
```

Example

Here, we are finding the Hamiltonian of spins interacting with homogeneous magnetic field in Y direction.

```fortran
program example
implicit none
integer,parameter::s=3
integer::i
complex*16::H(0:2**s-1,0:2**s-1)
real*8::B(0:s-1)
B=1.0d0
call YHAM(s,B,H)
do i=0,2**s-1,1
write(111,*)int(aimag(H(i,:)))
end do
end program
```

The file "fort.111" contains the matrix elements of the required Hamiltonian.

0	-1	-1	0	-1	0	0	0
1	0	0	-1	0	-1	0	0
1	0	0	-1	0	0	-1	0
0	1	1	0	0	0	0	-1
1	0	0	0	0	-1	-1	0
0	1	0	0	1	0	0	-1
0	0	1	0	1	0	0	-1
0	0	0	1	0	1	1	0

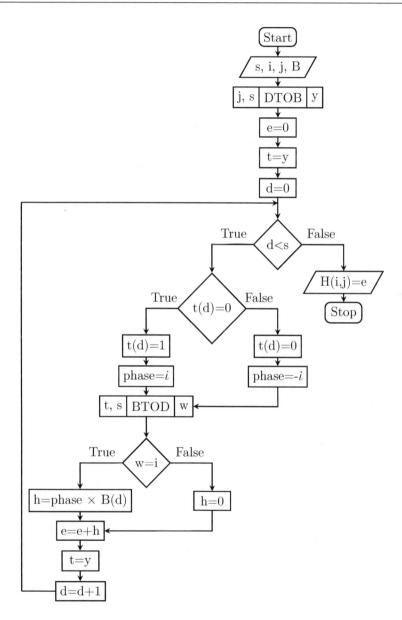

FIGURE 6.2: Flowchart of YHAM

6.1.3 Hamiltonian of spins interacting with an external magnetic field in the Z direction

In this case we choose $j = z$ in Eq. [6.6] and thereby, the interaction Hamiltonian of a linear array of spins interacting with the external magnetic field in the z direction is given by,

$$H = \sum_{i=1}^{N} B_i \sigma_i^z. \tag{6.9}$$

---

subroutine ZHAM(s,B,HB)

Parameters

In/Out	Argument	Description
[in]	s	s (integer) is the total number of spins.
[in]	B	B is a real*8 array of dimension (0:s−1). The ith entry of this array is the strength of the applied magnetic field along Z direction on the ith spin
[out]	HB	HB is a real*8 array of dimension (0:2**s−1,0:2**s−1) which is the required Hamiltonian

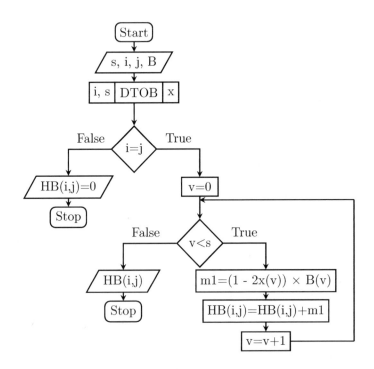

FIGURE 6.3: Flowchart of ZHAM

Implementation

```
subroutine ZHAM(s,B,HB)
implicit none
integer::s,i,j,v,x(0:s-1)
real*8::B(0:s-1)
real*8::HB(0:2**s-1,0:2**s-1),m1
HB=0.0d0
do i=0,2**s-1,1
    do j=i,2**s-1,1
        call DTOB(i,x,s)
        if (i==j) then
            do v=0,s-1
                m1=(1-2*x(v))*B(v)
                HB(i,j)=HB(i,j)+m1
            end do
        else
            HB(i,j)=0
        end if
    end do
end do
end subroutine
```

Example

Here, we are finding the Hamiltonian of spins interacting with inhomogeneous magnetic field in Z direction.

```
program example
implicit none
integer,parameter::s=3
integer::i
real*8::H(0:2**s-1,0:2**s-1)
real*8::B(0:s-1)
B=(/1.0d0,2.0d0,3.0d0/)
call ZHAM(s,B,H)
do i=0,2**s-1,1
    write(111,*)int(H(i,:))
end do
end program
```

The file "fort.111" contains the matrix elements of required Hamiltonian.

6	0	0	0	0	0	0	0
0	0	0	0	0	0	0	0
0	0	2	0	0	0	0	0
0	0	0	-4	0	0	0	0
0	0	0	0	4	0	0	0
0	0	0	0	0	-2	0	0
0	0	0	0	0	0	0	0
0	0	0	0	0	0	0	-6

6.2   Hamiltonian for the direct exchange spin-spin interaction

The Hamiltonian for the quantum mechanical spin-spin exchange interaction is given by:

$$H = \sum_{i=1}^{N} J_r^i \sigma_i^j \sigma_{i+r}^j. \tag{6.10}$$

Here J_r^i is the interaction strength (which can be both positive and negative) which is also called the coupling constant between the ith and the $(i+r)$th spins with r representing the rth neighboring interaction. Here N is the total number of spins (qubits) present in the system and σ^j are the Pauli spin matrices, where $j = x$, y, z. The notation σ_{i+r}^j denotes the Pauli spin matrices corresponding to the rth neighboring spin. Note that J_r^i may be a constant or it can be chosen from any probability distribution to induce bond randomness. In what follows below, we discuss three cases.

6.2.1   Spin-spin interaction in X direction

The Hamiltonian for the spin-spin interaction in the X direction is obtained by, choosing $j = x$ in Eq. [6.10] as

$$H = \sum_{i=1}^{N} J_r^i \sigma_i^x \sigma_{i+r}^x. \tag{6.11}$$

---

subroutine ISINGXHAM(s,r,BC,Jr,HI)

Parameters

In/Out	Argument	Description
[in]	s	s (integer) is the total number of spins
[in]	r	r (integer) is the rth nearest neighbour interaction
[in]	BC	BC (character*1) is a string defining the type of boundary condition if BC = 'y', periodic boundary conditions will be considered if BC = 'n', open boundary conditions will be considered
[in]	Jr	Jr is a real*8 array of dimension (0:s−1). Here ith entry of this array represents the interaction strength between the ith and (i+r)th spin
[out]	HI	HI is a real*8 array of dimension (0:2**s−1,0:2**s−1) which is the required Hamiltonian

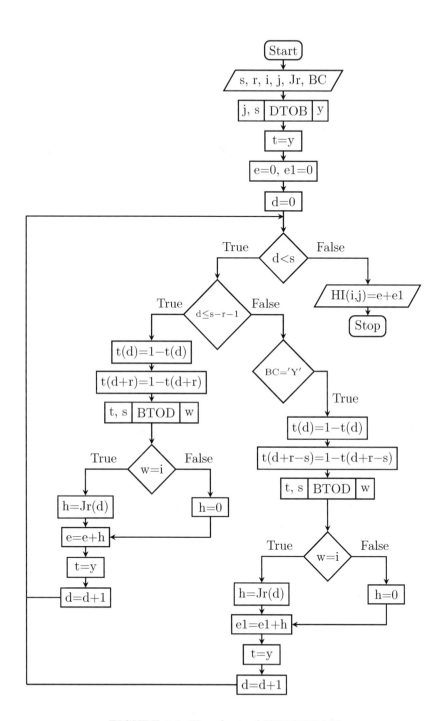

FIGURE 6.4: Flowchart of ISINGXHAM

Implementation

```
subroutine ISINGXHAM(s,r,BC,Jr,HI)
implicit none
integer::s,r,t(0:s-1),y(0:s-1),i,j,d,w
real*8::HI(0:2**s-1,0:2**s-1),e,e1
real*8::h,Jr(0:s-1)
character*1::BC
if (r>s) then
    print*, 'r should be less than s'
    stop
end if
HI=0.0d0
do i=0,2**s-1,1
    do j=i,2**s-1,1
        call DTOB(j,y,s)
        e=0
        e1=0
        t=y
        do d=0,s-1
            if(d.le.s-r-1) then
                t(d)=1-t(d)
                t(d+r)=1-t(d+r)
                call BTOD(t,w,s)
                if(w==i) then
                    h=Jr(d)
                else
                    h=0
                end if
                e=e+h
                t=y
            else
                if (BC=='y' ) then
                    t(d)=1-t(d)
                    t(d+r-s)=1-t(d+r-s)
                    call BTOD(t,w,s)
                    if(w==i) then
                        h=Jr(d)
                    else
                        h=0
                    end if
                    e1=e1+h
                    t=y
                end if
            end if
        end do
        HI(i,j)=e+e1
        HI(j,i)=HI(i,j)
    end do
end do
end subroutine
```

Example

In this example, we are finding the Hamiltonian of spin-spin isotropic interaction (Jr=(1,1,1)) with periodic boundary conditions in X direction.

```fortran
program example
implicit none
integer,parameter::s=3
integer::i
real*8::H(0:2**s-1,0:2**s-1),random(0:s-1)
real*8::Jr(0:s-1)
Jr=1.0d0
call ISINGXHAM(s,1,'y',Jr,H)
do i=0,2**s-1,1
   write(111,*)int(H(i,:))
end do
end program
```

The file "fort.111" contains the matrix elements of the required Hamiltonian.

0	0	0	1	0	1	1	0
0	0	1	0	1	0	0	1
0	1	0	0	1	0	0	1
1	0	0	0	0	1	1	0
0	1	1	0	0	0	0	1
1	0	0	1	0	0	1	0
1	0	0	1	0	1	0	0
0	1	1	0	1	0	0	0

6.2.2   Spin-spin interaction in Y direction

The Hamiltonian for the spin-spin exchange interaction in the Y direction is obtained by, choosing $j = y$ in Eq. [6.10] as

$$H = \sum_{i=1}^{N} J_r^i \sigma_i^y \sigma_{i+r}^y. \tag{6.12}$$

```fortran
subroutine ISINGYHAM(s,r,BC,Jr,HI)
```

Parameters

In/Out	Argument	Description
[in]	s	s (integer) is the total number of spins
[in]	r	r (integer) is the rth nearest neighbour interaction

In/Out	Argument	Description
[in]	BC	BC (character*1) is a string defining the type of boundary condition if BC = 'y', periodic boundary conditions will be considered if BC = 'n', open boundary conditions will be considered
[in]	Jr	Jr is a real*8 array of dimension $(0{:}s-1)$. Here ith entry of this array represents the interaction strength between the ith and (i+r)th spin
[out]	HI	HI is a real*8 array of dimension $(0{:}2^{**}s-1,0{:}2^{**}s-1)$ which is the required Hamiltonian

Implementation

```fortran
subroutine ISINGYHAM(s,r,BC,Jr,HI)
implicit none
integer*4::s,r,t(0:s-1),y(0:s-1),d,w,i,j
real*8::HI(0:2**s-1,0:2**s-1),e,e1,Jr(0:s-1)
complex*16::i2,p1,p2,p3,p4,h
character*1::BC
if (r>s) then
    print*, 'r should be less than s'
    stop
end if
HI=0.0d0
i2=dcmplx(0,1)
do i=0,2**s-1,1
    do j=i,2**s-1,1
        call DTOB(j,y,s)
        e=0
        e1=0
        t=y
        do d=0,s-1
            if(d.le.s-r-1) then
                p1=i2*((-1)**(t(d)))
                p2=i2*((-1)**(t(d+r)))
                t(d)=1-t(d)
                t(d+r)=1-t(d+r)
                call BTOD(t,w,s)
                if(w==i) then
                    h=p1*p2*Jr(d)
                else
                    h=0
                end if
                e=e+real(h)
                t=y
            else
                if (BC=='y') then
```

```
                    p3=i2*((-1)**(t(d)))
                    p4=i2*((-1)**(t(d+r-s)))
                    t(d)=1-t(d)
                    t(d+r-s)=1-t(d+r-s)
                    call BTOD(t,w,s)
                    if(w==i) then
                        h=p3*p4*Jr(d)
                    else
                        h=0
                    end if
                    e1=e1+real(h)
                    t=y
                end if
            end if
        end do
        HI(i,j)=e+e1
        HI(j,i)=HI(i,j)
    end do
end do
end subroutine
```

Example

In this example, we are finding the Hamiltonian of spin-spin random interaction (Jr=(1,2,3)) with open boundary conditions in Y direction.

```
program example
implicit none
integer,parameter::s=3
integer::i
real*8::H(0:2**s-1,0:2**s-1)
real*8::Jr(0:s-1)
Jr=(/1.0d0, 2.0d0, 3.0d0/)
call ISINGYHAM(s,1,'n',Jr,H)
do i=0,2**s-1,1
    write(111,*)int(H(i,:))
end do
end program
```

The file "fort.111" contains the matrix elements of the required Hamiltonian.

0	0	0	-2	0	0	-1	0
0	0	2	0	0	0	0	-1
0	2	0	0	1	0	0	0
-2	0	0	0	0	1	0	0
0	0	1	0	0	0	0	-2
0	0	0	1	0	0	2	0
-1	0	0	0	0	2	0	0
0	-1	0	0	-2	0	0	0

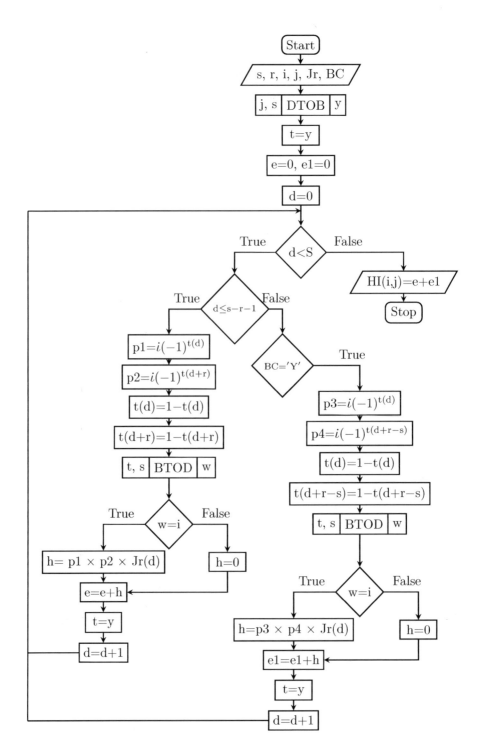

FIGURE 6.5: Flowchart of ISINGYHAM

6.2.3 Spin-spin interaction in Z direction

The Hamiltonian for the spin-spin exchange interaction in the Z direction is obtained by, choosing $j = z$ in Eq. [6.10] as

$$H = \sum_{i=1}^{N} J_r^i \sigma_i^z \sigma_{i+r}^z. \tag{6.13}$$

```
subroutine ISINGZHAM(s,r,BC,Jr,HI)
```

Parameters

In/Out	Argument	Description
[in]	s	s (integer) is the total number of spins
[in]	r	r (integer) is the rth nearest neighbour interaction
[in]	BC	BC (character*1) is a string defining the type of boundary condition if BC = 'y', periodic boundary conditions will be considered if BC = 'n', open boundary conditions will be considered
[in]	Jr	Jr is a real*8 array of dimension (0:s−1). Here ith entry of this array represents the interaction strength between the ith and (i+r)th spin
[out]	HI	HI is a real*8 array of dimension (0:2**s−1,0:2**s−1) which is the required Hamiltonian

Implementation

```
subroutine ISINGZHAM(s,r,BC,Jr,HI)
implicit none
integer*4::s,r,i,j,v,x(0:s-1)
real*8::HI(0:2**s-1,0:2**s-1),m1,Jr(0:s-1)
character*1::BC
if (r>s) then
    print*, 'r should be less than s'
    stop
end if
HI=0.0d0
do i=0,2**s-1,1
    do j=i,2**s-1,1
        HI(i,j)=0.0d0
        call DTOB(j,x,s)
        if (i==j) then
            do v=0,s-1
                if(v.le.s-r-1) then
                    m1=(1-2*x(v))*(1-2*x(v+r))*Jr(v)
                    HI(i,j)=HI(i,j)+m1
                else
```

```
            if (BC=='y') then
                m1=(1-2*x(v))*(1-2*x(v+r-s))*Jr(v)
                HI(i,j)=HI(i,j)+m1
            end if
        end if
    end do
end if
end do
end do
end subroutine
```

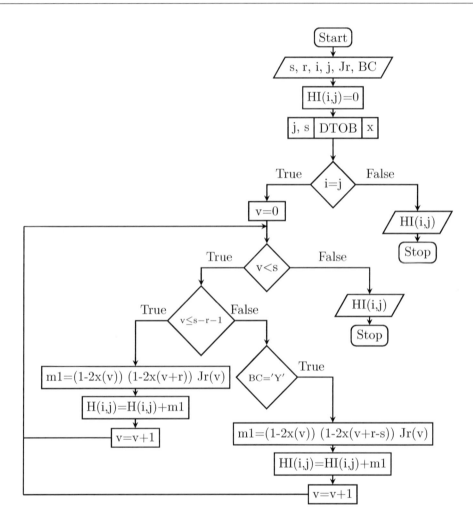

FIGURE 6.6: Flowchart of ISINGZHAM

Example

In this example, we are finding the Hamiltonian of spin-spin isotropic interaction (Jr=(1,1,1)) with periodic boundary conditions in Z direction.

```
program example
```

```
implicit none
integer,parameter::s=3
integer::i
real*8::H(0:2**s-1,0:2**s-1)
real*8::Jr(0:s-1)
Jr=1.0d0
call ISINGZHAM(s,1,'y',Jr,H)
do i=0,2**s-1,1
    write(111,*)int(H(i,:))
end do
end program
```

The file "fort.111" contains the matrix elements of the required Hamiltonian.

3	0	0	0	0	0	0	0
0	-1	0	0	0	0	0	0
0	0	-1	0	0	0	0	0
0	0	0	-1	0	0	0	0
0	0	0	0	-1	0	0	0
0	0	0	0	0	-1	0	0
0	0	0	0	0	0	-1	0
0	0	0	0	0	0	0	3

6.2.4 Variants of the Ising model

Now we will see several variants of the Ising model which is being used in condensed matter physics.

6.2.4.1 The isotropic Ising model in a non-zero uniform magnetic field

The Hamiltonian of Ising model is given by,

$$H = -J \sum_{i=1}^{N} \sigma_i^z \sigma_{i+1}^z - B \sum_{i=1}^{N} \sigma_i^z. \tag{6.14}$$

Using our recipes we can construct the above Hamiltonian as follows. We have taken the example of $N = 3$, $J = 1$, $B = 1$ and $r = 1$ under PBC.

```
program example
implicit none
integer,parameter::s=3
integer::i
real*8,dimension(0:2**s-1,0:2**s-1)::H1,H2,H
real*8::Jr(0:s-1),B(0:s-1)
Jr=1.0d0
B=1.0d0
call ISINGZHAM(s,1,'y',Jr,H1)
call ZHAM(s,B,H2)
H=-H1-H2
end program
```

In the above program, the variable H contains the required Hamiltonian matrix as shown in Eq. [6.14]

6.2.4.2 The isotropic transverse Ising model in a non-zero uniform magnetic field

The Hamiltonian of the transverse field Ising model [104] is given by,

$$H = -J\sum_{i=1}^{N}\sigma_i^z\sigma_{i+1}^z - B\sum_{i=1}^{N}\sigma_i^x \qquad (6.15)$$

Using our recipes we can construct the above Hamiltonian as follows. We have taken the example of $N = 3$, $J = 1$, $B = 1$, and $r = 1$ under PBC.

```
program example
implicit none
integer,parameter::s=3
integer::i
real*8,dimension(0:2**s-1,0:2**s-1)::H,H1,H2
call ISINGZHAM(s,1,'y',1.0d0,H1)
call XHAM(s,1.0d0,H2)
H=-H1-H2
end program
```

In the above program, the variable H contains the required Hamiltonian matrix as shown in Eq. [6.15]. The tilted field Ising model as well as other Ising type models with any neighbor interaction can simply be constructed suitably using similar prescriptions as given in sections [6.2.4.1, 6.2.4.2].

6.3 The Heisenberg interaction

The one-dimensional Heisenberg model [93, 105–107] is an important model in physics, The Hamiltonian for the nearest-neighbor one-dimensional Heisenberg model for the case of N interacting spin half qubits is as follows

$$H = J\sum_{i=1}^{N}\vec{\sigma}_i.\vec{\sigma}_{i+1} = J\sum_{i=1}^{N}(\sigma_i^x\sigma_{i+1}^x + \sigma_i^y\sigma_{i+1}^y + \sigma_i^z\sigma_{i+1}^z) \qquad (6.16)$$

Note the dot product in the above Hamiltonian. If $J < 0$, the ground state of the above Hamiltonian is ferromagnetic and If $J > 0$, the ground state is antiferromagnetic. It can be solved analytically for its ground states as well as the excited states by Bethe Ansatz [105] for the case when $J < 0$ and by Hulthen's method [93] for the case when $J > 0$. We generalize the Heisenberg Hamiltonian to the rth neighbor with random interaction strengths using Eq. [6.16] as,

$$H = \sum_{i=1}^{N}J_r^i\vec{\sigma}_i.\vec{\sigma}_{i+r} = \sum_{i=1}^{N}J_r^i(\sigma_i^x\sigma_{i+r}^x + \sigma_i^y\sigma_{i+r}^y + \sigma_i^z\sigma_{i+r}^z), \qquad (6.17)$$

where J_r^i is the interaction strength which is also called the coupling constant between the the ith and $(i + r)$th spin. $\vec{\sigma} = (\sigma^x, \sigma^y, \sigma^z)$ where σ^j are the Pauli spin matrices, with $j = x$, y, z. Also note, σ_{i+r} denotes the Pauli spin matrices corresponding to the rth neighboring spin from σ_i. It is very easy to verify that,

$$\vec{\sigma}_i.\vec{\sigma}_{i+r} = 2P_{i,i+r} - 1 \qquad (6.18)$$

Where $P_{i,i+r}$ is the permutation operator which swaps the ith and $(i+r)$th bit. We use the permutation operator to construct the Hamiltonian instead of using the naive form $\vec{\sigma}_i.\vec{\sigma}_{i+r}$.

```
subroutine HEIHAM(s,r,BC,Jr,H)
```

Parameters

In/Out	Argument	Description
[in]	s	s (integer) is the total number of spins
[in]	r	r (integer) is the rth nearest neighbor interaction
[in]	BC	BC (character*1) is a string defining the type of boundary condition if BC = 'y', periodic boundary conditions will be considered if BC = 'n', open boundary conditions will be considered
[in]	Jr	Jr is a real*8 array of dimension (0:s−1). Here ith entry of this array represents the interaction strength between the ith and (i+r)th spin
[out]	H	H is a real*8 array of dimension (0:2**s−1,0:2**s−1) which is the required Hamiltonian

Implementation

```
subroutine HEIHAM(s,r,BC,Jr,H)
implicit none
integer::s,r,i,j,y(0:s-1),z(0:s-1),o,w,temp
real*8,dimension(0:2**s-1,0:2**s-1)::H
real*8::c,h1,h2,h3,h4,h5
character*1::BC
real*8::Jr(0:s-1),summ
summ=0.0d0
if (BC=='y') then
    do i=0,s-1
        summ=summ+Jr(i)
    end do
end if
if (BC=='n') then
    do i=0,s-r-1
        summ=summ+Jr(i)
    end do
end if
do i=0,2**s-1,1
    do j=i,2**s-1,1
        call DTOB(j,y,s)
        h1=0
        h2=0
        h3=0
```

```fortran
      h4=0
      do o=0,s-1,1
         z=y
         if(o.le.s-r-1) then
            temp=z(o)
            z(o)=z(o+r)
            z(o+r)=temp
            call BTOD(z,w,s)
            if(w==i) then
               c=Jr(o)
            else
               c=0
            end if
            h1=h1+c
         else
            if (BC=='y') then
               temp=z(o)
               z(o)=z(o+r-s)
               z(o+r-s)=temp
               call BTOD(z,w,s)
               if(w==i) then
                  c=Jr(o)
               else
                  c=0
               end if
               h2=h2+c
            end if
         end if
      end do
      if(i==j)then
         h3=summ
      else
         h3=0
      end if
      h4=h1+h2
      H(i,j)=4.0d0*(h4/2.0d0-h3/4.0d0)
      H(j,i)=H(i,j)
   end do
end do
end subroutine
```

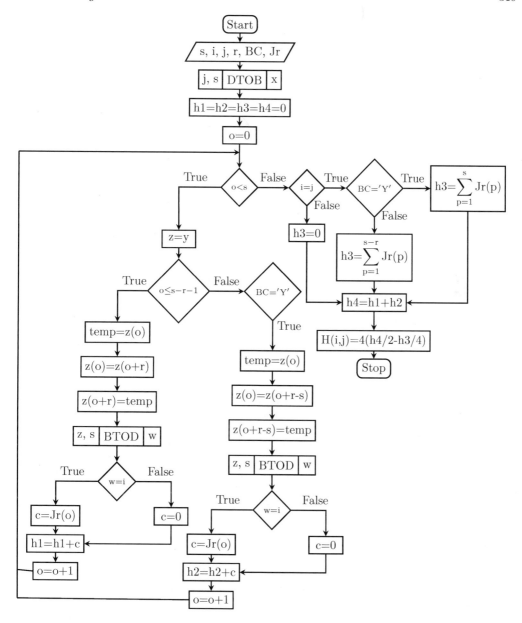

FIGURE 6.7: Flowchart of HEIHAM

Example

In this example we are finding the Hamiltonian of the Heisenberg random interaction (Jr=(1,2,3)) with periodic boundary conditions.

```
program example
implicit none
integer,parameter::s=3
integer::i
real*8::H(0:2**s-1,0:2**s-1),Jr(0:s-1)
```

```
Jr=(/1.0d0,2.0d0,3.0d0/)
call HEIHAM(s,1,'y',Jr,H)
do i=0,2**s-1
write(111,*)int(H(i,:))
end do
end program
```

The file "fort.111" contains the matrix elements of the required Hamiltonian.

3	0	0	0	0	0	0	0
0	-4	4	0	6	0	0	0
0	4	0	0	2	0	0	0
0	0	0	-2	0	2	6	0
0	6	2	0	-2	0	0	0
0	0	0	2	0	0	4	0
0	0	0	6	0	4	-4	0
0	0	0	0	0	0	0	6

6.3.1 Variants of the Heisenberg model

The Hamiltonian described by Eq. [6.16] and Eq. [6.17] represent the Heisenberg XXX model. Of course a generic Heisenberg model with nearest-neighbor interaction will be of the form,

$$H = \sum_{i=1}^{N} (J_{rx}\sigma_i^x\sigma_{i+r}^x + J_{ry}\sigma_i^y\sigma_{i+r}^y + J_{rz}\sigma_i^z\sigma_{i+r}^z), \qquad (6.19)$$

here if $J_{rx} = J_{ry} = J_{rz}$, we call it the XXX model, if $J_{rx} \neq J_{ry} \neq J_{rz}$ then it is called the XYZ model and if $J = J_{rx} = J_{ry} \neq J_{rz}$, it is called the XXZ model. These can be readily generalized to the rth neighbor models. Note that here J_{rj} does not contain superscript i as the original Heisenberg model does not have bond randomness.

6.3.1.1 The XXX model

Using our recipes we can construct the XXX Hamiltonian as follows. We have taken the example of $N = 3$, $J_{rx} = J_{ry} = J_{rz} = 1$ and $r = 1$ under PBC.

```
program example
implicit none
integer,parameter::s=3
integer::i
real*8,dimension(0:2**s-1,0:2**s-1)::H,H1,H2,H3
real*8::Jr(0:s-1)
Jr=1.0d0
call ISINGXHAM(s,1,'y',Jr,H1)
call ISINGYHAM(s,1,'y',Jr,H2)
call ISINGZHAM(s,1,'y',Jr,H3)
H=H1+H2+H3
end program
```

In the above program, the variable H contains the required Hamiltonian matrix.

6.3.1.2   The XYZ model

Using our recipes we can construct the XYZ Hamiltonian as follows. We have taken the example of $N = 3$, $J_{rx} = 0.7$, $J_{ry} = 0.8$, $J_{rz} = 0.9$ and $r = 1$ under PBC.

```
program example
implicit none
integer,parameter::s=3
integer::i
real*8,dimension(0:2**s-1,0:2**s-1)::H,H1,H2,H3
real*8::Jr1(0:s-1),Jr2(0:s-1),Jr3(0:s-1)
Jr1=0.70d0
Jr2=0.8d0
Jr3=0.9d0
call ISINGXHAM(s,1,'y',Jr1,H1)
call ISINGYHAM(s,1,'y',Jr2,H2)
call ISINGZHAM(s,1,'y',Jr3,H3)
H=H1+H2+H3
end program
```

In the above program, the variable H contains the required Hamiltonian matrix.

6.3.1.3   The Majumdar-Ghosh model

Hamiltonian of Majumdar-Ghosh model [108–110] is given by,

$$H = J_1 \sum_{i=1}^{N} \vec{\sigma}_i . \vec{\sigma}_{i+1} + J_2 \sum_{i=1}^{N} \vec{\sigma}_i . \vec{\sigma}_{i+2} \tag{6.20}$$

Using our recipes we can construct the Majumdar-Ghosh Hamiltonian as follows. We have taken the example of $N = 4$, $J_1 = 1$, $J_2 = 0.5$ and $r = 1, 2$ under PBC. Note that J_1 and J_2 are isotropic.

```
program example
implicit none
integer,parameter::s=4
integer::i
real*8,dimension(0:2**s-1,0:2**s-1)::H1,H2,H
real*8::Jr1(0:s-1),Jr2(0:s-1)
Jr1=1.0d0
Jr2=0.5d0
call HEIHAM(s,1,'y',Jr1,H1)
call HEIHAM(s,2,'y',Jr2,H2)
H=H1+H2
end program
```

In the above program, the variable H contains the required Hamiltonian matrix.

6.4   The Dzyaloshinskii-Moriya interaction

The magnetic interactions between magnetic ions in a solid depend on numerous factors like neighboring ions, temperature, external fields, etc. In some cases to describe the system one uses Hamiltonian involving simultaneous interaction between several spins also. Apart from the direct exchange coupling involving the dot product of the nearest neighbor spin operators $\vec{\sigma}_i.\vec{\sigma}_{i+1}$ in the Hamiltonian as discussed earlier in section [6.3], however, there is a small antisymmetric part due to the $L-S$ coupling between the spins or ions at site i and $i+1$. This favors spin canting (a tilt between the spins near to zero temperature) which we call an indirect coupling. It is interesting to note that this exists in a number of ionic solids, by which the short ranged exchange interaction between two non-neighboring magnetic ions is mediated by a non-magnetic ion that resides between them in a phenomena similar to the Anderson super exchange [111]. This indirect coupling part of the Hamiltonian is called the Dzyaloshinskii-Moriya (DM) interaction [112, 113]. The Hamiltonian of the DM interaction is given by,

$$H = \sum_{i=1}^{N} \vec{D}.\vec{\sigma}_i \times \vec{\sigma}_{i+1}, \tag{6.21}$$

where $\vec{D} = (D_x, D_y, D_z)$ is the DM vector, whose components specify the coupling parameter of the DM interaction in the X, Y, Z directions, respectively, N is the total number of spins or qubits. Note that for the DM interaction between spins at sites i and $i+1$, the vector D is orthogonal to both the spins. That is, if the two interacting spins are lying in the plane of this book, then the DM vector is along the direction either into or out of the plane containing book. This cross product term as given in Eq. [6.21] is also related to what is known as the Heisenberg spin current [114]. Since the interaction is antisymmetric, the above Hamiltonian involves a cross product. We can recast Eq. [6.21] as,

$$H = \sum_{i=1}^{N} \begin{vmatrix} D_x & D_y & D_z \\ \sigma_i^x & \sigma_i^y & \sigma_i^z \\ \sigma_{i+1}^x & \sigma_{i+1}^y & \sigma_{i+1}^z \end{vmatrix}. \tag{6.22}$$

By expanding the determinant in equation Eq. [6.22] we get,

$$H = \sum_{i=1}^{N} \Big[D_x(\sigma_i^y\sigma_{i+1}^z - \sigma_i^z\sigma_{i+1}^y) + D_y(\sigma_i^z\sigma_{i+1}^x - \sigma_i^x\sigma_{i+1}^z)$$
$$+ D_z(\sigma_i^x\sigma_{i+1}^y - \sigma_i^y\sigma_{i+1}^x) \Big]. \tag{6.23}$$

We give the recipes for constructing these three terms in the Hamiltonian separately. The above formalism of antisymmetric exchange can be faithfully extended to rth nearest neighbors also and in fact the recipes we give involve DM interaction between the spin at the site i and the spin at site the $i+r$ as

$$H = \sum_{i=1}^{N} \Big[D_{rx}^i(\sigma_i^y\sigma_{i+r}^z - \sigma_i^z\sigma_{i+r}^y) + D_{ry}^i(\sigma_i^z\sigma_{i+r}^x - \sigma_i^x\sigma_{i+r}^z)$$
$$+ D_{rz}^i(\sigma_i^x\sigma_{i+r}^y - \sigma_i^y\sigma_{i+r}^x) \Big]. \tag{6.24}$$

Also note that to accommodate the random couplings between ith and $(i+r)$th spins, we make the components of $\vec{D}$ random by introducing the superscript i. It is noteworthy to mention that, in general, in Eq. [6.23], only one component of the vector D is non-zero, depending on the direction of the orientation of vector D.

6.4.1 DM vector in X direction

This corresponds to the first term in Eq. [6.24] which is,

$$H = \sum_{i=1}^{N} D_{rx}^{i}(\sigma_i^y \sigma_{i+r}^z - \sigma_i^z \sigma_{i+r}^y). \tag{6.25}$$

subroutine DMHAMX(s,r,BC,Dr,HDM)

Parameters

In/Out	Argument	Description
[in]	s	s (integer) is the total number of spins
[in]	r	r (integer) is the rth nearest neighbor interaction
[in]	BC	BC (character*1) is a string defining the type of boundary condition if BC = 'y', periodic boundary conditions will be considered if BC = 'n', open boundary conditions will be considered
[in]	Dr	Dr is a real*8 array of dimension (0:s−1). Here ith entry of this array represents X component of the DM vector coupling the ith and $(i+r)$th spin
[out]	HDM	HDM is a real*8 array of dimension (0:2**s−1,0:2**s−1) which is the required Hamiltonian

Implementation

```
subroutine DMHAMX(s,r,BC,Dr,HDM)
implicit none
integer::s,r
integer::x,i,j,t(0:s-1),y(0:s-1),w1
real*8::Dr(0:s-1),m
character*1::BC
complex*16::i2,phase,h,HDM(0:2**s-1,0:2**s-1),e,e1,h1,h2,e2,e3
i2=dcmplx(0,1)
do i=0,2**s-1
    do j=i,2**s-1
        call DTOB(j,y,s)
        e=0
        e1=0
        t=y
        e2=0
        e3=0
```

```
do x=0,s-1
    if(x.le.s-r-1) then
        phase=i2*((-1)**(t(x)))
        t(x)=1-t(x)
        m=1-2*t(x+r)
        call BTOD(t,w1,s)
        if(w1==i) then
            h=m*phase*Dr(x)
        else
            h=0
        end if
        e=e+h
        t=y
    else
        if (BC=='y') then
            phase=i2*((-1.0)**(t(x)))
            t(x)=1-t(x)
            m=1-2*t(x+r-s)
            call BTOD(t,w1,s)
            if(w1==i) then
                h=m*phase*Dr(x)
            else
                h=0
            end if
            e1=e1+h
            t=y
        end if
    end if
end do
h1=e+e1
do x=0,s-1
    if(x.le.s-r-1) then
        phase=i2*((-1.0)**(t(x+r)))
        m=1-2*t(x)
        t(x+r)=1-t(x+r)
        call BTOD(t,w1,s)
        if(w1==i) then
            h=m*phase*Dr(x)
        else
            h=0
        end if
        e2=e2+h
        t=y
    else
        if (BC=='y') then
            phase=i2*((-1)**(t(x+r-s)))
            t(x+r-s)=1-t(x+r-s)
            m=1-2*t(x)
            call BTOD(t,w1,s)
            if(w1==i) then
                h=m*phase*Dr(x)
```

```
            else
                h=0
            end if
            e3=e3+h
            t=y
        end if
      end if
    end do
    h2=e2+e3
    HDM(i,j)=h1-h2
    HDM(j,i)=conjg(HDM(i,j))
  end do
end do
end subroutine
```

Example

In this example we are finding the DM Hamiltonian with random interaction couplings (Dr=(1,2,3)) in the X direction with periodic boundary condition.

```
program example
implicit none
integer,parameter::s=3
integer::i
complex*16::H(0:2**s-1,0:2**s-1)
real*8::Dr(0:s-1)
Dr=(/1.0d0,2.0d0,3.0d0/)
call DMHAMX(s,1,'y',Dr,H)
do i=0,2**s-1,1
write(111,*)int(aimag(H(i,:)))
end do
end program
```

The file "fort.111" contains the matrix elements of the required Hamiltonian.

0	-1	-1	0	2	0	0	0
1	0	0	3	0	-4	0	0
1	0	0	-5	0	0	4	0
0	-3	5	0	0	0	0	-2
-2	0	0	0	0	5	-3	0
0	4	0	0	-5	0	0	1
0	0	-4	0	3	0	0	1
0	0	0	2	0	-1	-1	0

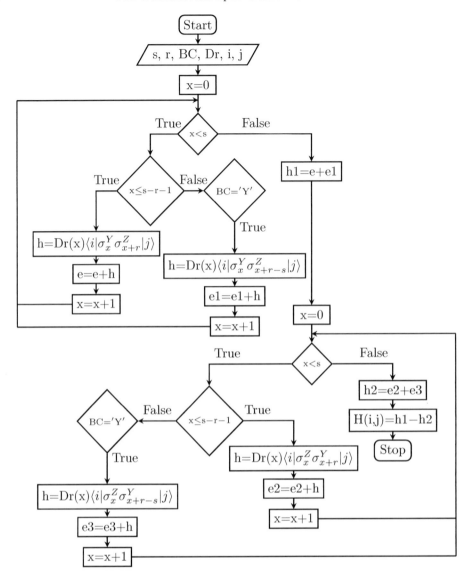

FIGURE 6.8: Flowchart of DMHAMX

6.4.2 DM vector in Y direction

This corresponds to the second term in Eq. [6.24] which is,

$$H = \sum_{i=1}^{N} D_{ry}^{i} (\sigma_i^z \sigma_{i+r}^x - \sigma_i^x \sigma_{i+r}^z).$$ (6.26)

```
subroutine DMHAMY(s,r,BC,Dr,HDM)
```

Parameters

In/Out	Argument	Description
[in]	s	s (integer) is the total number of spins
[in]	r	r (integer) is the rth nearest neighbor interaction
[in]	BC	BC (character*1) is a string defining the type of boundary condition if BC = 'y', periodic boundary conditions will be considered if BC = 'n', open boundary conditions will be considered
[in]	Dr	Dr is a real*8 array of dimension (0:s−1). Here ith entry of this array represents Y component of the DM vector coupling the ith and $(i+r)$th spin
[out]	HDM	HDM is a real*8 array of dimension (0:2**s−1,0:2**s−1) which is the required Hamiltonian

Implementation

```
subroutine DMHAMY(s,r,BC,Dr,HDM)
implicit none
integer::s,r
integer::x,i,j,t(0:s-1),y(0:s-1),w1
real*8::Dr(0:s-1),m
character*1::BC
real*8::i2,h,HDM(0:2**s-1,0:2**s-1),e,e1,h1,h2,e2,e3
do i=0,2**s-1
    do j=i,2**s-1
        call DTOB(j,y,s)
        e=0
        e1=0
        t=y
        e2=0
        e3=0
        do x=0,s-1
            if(x.le.s-r-1) then
                t(x)=1-t(x)
                m=1-2*t(x+r)
                call BTOD(t,w1,s)
                if(w1==i) then
                    h=m*Dr(x)
                else
                    h=0
                end if
                e=e+h
                t=y
            else
                if (BC=='y') then
```

```
                    t(x)=1-t(x)
                    m=1-2*t(x+r-s)
                    call BTOD(t,w1,s)
                    if(w1==i) then
                         h=m*Dr(x)
                    else
                          h=0
                    end if
                    e1=e1+h
                    t=y
               end if
          end if
     end do
     h1=e+e1
     do x=0,s-1
          if(x.le.s-r-1) then
               m=1-2*t(x)
               t(x+r)=1-t(x+r)
               call BTOD(t,w1,s)
               if(w1==i) then
                    h=m*Dr(x)
               else
                    h=0
               end if
               e2=e2+h
               t=y
          else
               if (BC=='y') then
                    t(x+r-s)=1-t(x+r-s)
                    m=1-2*t(x)
                    call BTOD(t,w1,s)
                    if(w1==i) then
                         h=m*Dr(x)
                    else
                          h=0
                    end if
                    e3=e3+h
                    t=y
               end if
          end if
     end do
     h2=e2+e3
     HDM(i,j)=(h2-h1)
     HDM(j,i)=HDM(i,j)
  end do
end do
end subroutine
```

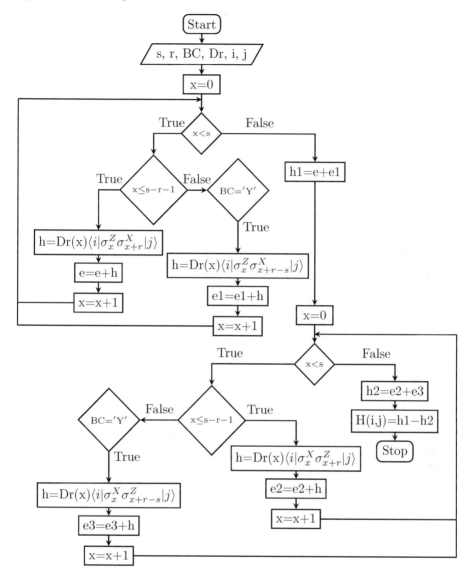

FIGURE 6.9: Flowchart of DMHAMY

Example

In this example we are finding the DM Hamiltonian with random interaction couplings (Dr=(1,1,1)) in the Y direction with periodic boundary condition.

```
program example
implicit none
integer,parameter::s=3
integer::i
real*8::H(0:2**s-1,0:2**s-1),Dr(0:s-1)
Dr=1.0d0
call DMHAMY(s,1,'y',Dr,H)
do i=0,2**s-1
```

```
write(111,*)int(H(i,:))
end do
end program
```

The file "fort.111" contains the matrix elements of the required Hamiltonian.

0	0	0	0	0	0	0	0
0	0	0	2	0	-2	0	0
0	0	0	-2	0	0	2	0
0	2	-2	0	0	0	0	0
0	0	0	0	0	2	-2	0
0	-2	0	0	2	0	0	0
0	0	2	0	-2	0	0	0
0	0	0	0	0	0	0	0

6.4.3   DM vector in Z direction

This corresponds to the third term in Eq. [6.24] which is,

$$H = \sum_{i=1}^{N} D_{rz}^{i}(\sigma_i^x \sigma_{i+r}^y - \sigma_i^y \sigma_{i+r}^x)). \tag{6.27}$$

```
subroutine DMHAMZ(s,r,BC,Dr,HDM)
```

Parameters

In/Out	Argument	Description
[in]	s	s (integer) is the total number of spins
[in]	r	r (integer) is the rth nearest neighbor interaction
[in]	BC	BC (character*1) is a string defining the type of boundary condition if BC = 'y', periodic boundary conditions will be considered if BC = 'n', open boundary conditions will be considered
[in]	Dr	Dr is a real*8 array of dimension (0:s−1). Here ith entry of this array represents Z component of the DM vector coupling the ith and (i+r)th spin
[out]	HDM	HDM is a real*8 array of dimension (0:2**s−1,0:2**s−1) which is the required Hamiltonian

Implementation

```
subroutine DMHAMZ(s,r,BC,Dr,HDM)
implicit none
integer::s,r
integer::x,i,j,t(0:s-1),y(0:s-1),w1
```

```fortran
real*8::Dr(0:s-1),m
character*1::BC
complex*16::i2,phase,h,HDM(0:2**s-1,0:2**s-1),e,e1,h1,h2,e2,e3
i2=dcmplx(0,1)
do i=0,2**s-1
    do j=i,2**s-1
        call DTOB(j,y,s)
        e=0
        e1=0
        t=y
        e2=0
        e3=0
        do x=0,s-1
            if(x.le.s-r-1) then
                phase=i2*((-1)**(t(x+r)))
                t(x)=1-t(x)
                t(x+r)=1-t(x+r)
                call BTOD(t,w1,s)
                if(w1==i) then
                    h=phase*Dr(x)
                else
                    h=0
                end if
                e=e+h
                t=y
            else
                if (BC=='y') then
                    phase=i2*((-1.0)**(t(x+r-s)))
                    t(x)=1-t(x)
                    t(x+r-s)=1-t(x+r-s)
                    call BTOD(t,w1,s)
                    if(w1==i) then
                        h=phase*Dr(x)
                    else
                        h=0
                    end if
                    e1=e1+h
                    t=y
                end if
            end if
        end do
        h1=e+e1
        do x=0,s-1
            if(x.le.s-r-1) then
                phase=i2*((-1.0)**(t(x)))
                t(x)=1-t(x)
                t(x+r)=1-t(x+r)
                call BTOD(t,w1,s)
                if(w1==i) then
                    h=phase*Dr(x)
                else
```

```
              h=0
          end if
          e2=e2+h
          t=y
      else
          if (BC=='y') then
              phase=i2*((-1)**(t(x)))
              t(x+r-s)=1-t(x+r-s)
              t(x)=1-t(x)
              call BTOD(t,w1,s)
              if(w1==i) then
                  h=phase*Dr(x)
              else
                  h=0
              end if
              e3=e3+h
              t=y
          end if
      end if
  end do
  h2=e2+e3
  HDM(i,j)=(h1-h2)
  HDM(j,i)=conjg(HDM(i,j))
  end do
end do
end subroutine
```

Example

In this example we are finding the DM Hamiltonian with random interaction couplings (Dr=(1,1,1)) in the Z direction with periodic boundary condition.

```
program example
implicit none
integer,parameter::s=3
integer::i
complex*16::H(0:2**s-1,0:2**s-1)
real*8::Dr(0:s-1)
Dr=1.0d0
call DMHAMZ(s,1,'y',Dr,H)
do i=0,2**s-1,1
write(111,*)int(aimag(H(i,:)))
end do
end program
```

The file "fort.111" contains the matrix elements of the required Hamiltonian.

0	0	0	0	0	0	0	0
0	0	2	0	-2	0	0	0
0	-2	0	0	2	0	0	0
0	0	0	0	0	2	-2	0
0	2	-2	0	0	0	0	0

0	0	0	-2	0	0	2	0
0	0	0	2	0	-2	0	0
0	0	0	0	0	0	0	0

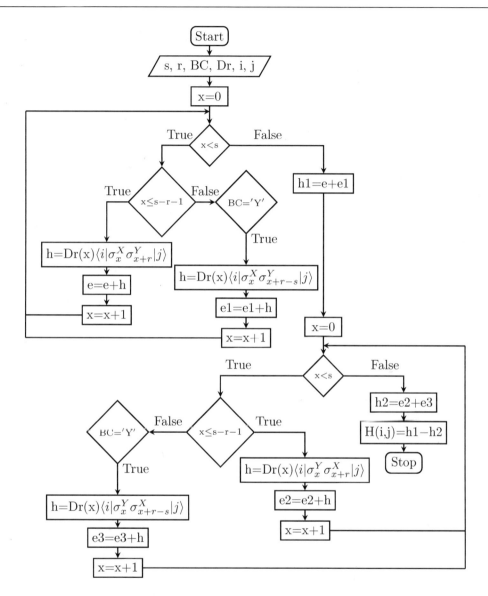

FIGURE 6.10: Flowchart of DMHAMZ

6.4.4   Illustrative example of constructing a combination Hamiltonian

Let us see how to construct a combination Hamiltonian which involves the Hamiltonian of the Heisenberg model with DM interaction is as follows. In the following example we have chosen the direction of the DM vector to be along Z.

$$H = J\sum_{i=1}^{N} \vec{\sigma}_i . \vec{\sigma}_{i+1} + D_z \sum_{i=1}^{N}(\sigma_i^x \sigma_{i+1}^y - \sigma_i^y \sigma_{i+1}^x) \qquad (6.28)$$

Using our recipes we can construct the above Hamiltonian as follows. We have taken the example of $N = 3$, $J = 1$, $D_z = 1$ and $r = 1$ under PBC.

```fortran
program example
implicit none
integer,parameter::s=3
integer::i
real*8,dimension(0:2**s-1,0:2**s-1)::H1
complex*16,dimension(0:2**s-1,0:2**s-1)::H,H2
real*8::Dr(0:s-1),Jr(0:s-1)
Dr=1.0d0
Jr=1.0d0
call HEIHAM(s,1,'y',Jr,H1)
call DMHAMZ(s,1,'y',Dr,H2)
H=H1+H2
end program
```

In the above program, the variable H contains the required Hamiltonian matrix.

7

Random Matrices and Random Vectors

In this chapter we give the numerical recipes to construct random numbers following a particular distribution, here we have taken the Gaussian and the uniform distribution. These are the most important distributions used in constructing random quantum states [115] as we progress through the chapter. More so, there are recipes to construct random matrices which occur in random matrix theory [116–118] and boderlining areas of quantum information. At the end of the chapter, recipes to construct the random density matrix is also given which will be very useful for people working on random quantum mixed states [119,120]. The examples indicating the output using these recipes are not given for a specific reason that, everytime the user prints an output, it will be different as they are from random processes.

7.1 Random number generator for Gaussian distribution

The probability density function for a Gaussian distribution is given by,

$$f(x) = \frac{1}{\sigma\sqrt{2\pi}}e^{-\frac{1}{2}\left(\frac{x-\mu}{\sigma}\right)^2} = \mathcal{N}(\mu, \sigma^2), \tag{7.1}$$

where μ, σ^2 are the mean and variance of the probability distribution, respectively. In FORTRAN, there is an inbuilt random number generator named as, CALL RANDOM_NUMBER() which generates an array of pseudo random numbers from the uniform distribution in the range $0 \leq x < 1$. Using these numbers we will generate random numbers corresponding to the Gaussian distribution $\mathcal{N}(\mu, \sigma^2)$. To achieve this, consider two uniformly distributed random numbers x_1 and x_2 along with the Box-Muller transformation [121] as given below:

$$Z = \mu + \sigma\sqrt{-2\log x_1}\cos 2\pi x_2. \tag{7.2}$$

Where Z is the Gaussian random variable with mean μ and standard deviation σ. It is important to note that, if $\mu = 0$ and $\sigma^2 = 1$, then the distribution is called the normal distribution given by $f_N(x) = e^{-\frac{x^2}{2}}/\sqrt{2\pi}$ and the notation for it is $\mathcal{N}(0, 1)$.

7.1.1 For a real random matrix whose elements are chosen from the Gaussian distribution

```
subroutine RANDDISTGAUSSR(d1,d2,mu,sigma,state)
```

Parameters

In/Out	Argument	Description
[in]	d1	d1 (integer) is the number of rows in the final matrix named as state
[in]	d2	d2 (integer) is the number of columns in the final matrix named as state
[in]	mu	mu (real*8) is the mean of the distribution
[in]	sigma	sigma (real*8) is the standard deviation of the distribution
[out]	state	state is a real*8 array of dimension (0:d1−1,0:d2−1) which is a random real matrix

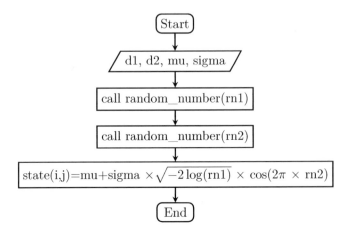

FIGURE 7.1: Flowchart of RANDDISTGAUSSR

Implementation

```
subroutine RANDDISTGAUSSR(d1,d2,mu,sigma,state)
implicit none
integer::d1,d2,i,j
real*8::state(d1,d2),rn1,rn2,pi,mu,sigma
pi=4.0d0*atan(1.0d0)
do i=1,d1
    do j=1,d2
        call random_number(rn1)
        call random_number(rn2)
        state(i,j)=mu+sigma*sqrt(-2*log(rn1))*cos(2*pi*rn2)
    end do
end do
end subroutine
```

Example

```
program example
implicit none
integer,parameter::d=5
integer::i,j
real*8::state(d,1)
call RANDDISTGAUSSR(d,1,0.0d0,1.d0,state)
end program
```

In the above example, state is a 5×1 real matrix which stores the random numbers from the Gaussian distribution with zero mean and unit standard deviation.

7.1.2 For a complex random matrix whose real and imaginary parts are individually chosen from the Gaussian distribution

```
subroutine RANDDISTGAUSSC(d1,d2,mu,sigma,state)
```

Parameters

In/Out	Argument	Description
[in]	d1	d1 (integer) is the number of rows in the final matrix named as state
[in]	d2	d2 (integer) is the number of columns in the final matrix named as state
[in]	mu	mu (real*8) is the mean of the distribution
[in]	sigma	sigma (real*8) is the standard deviation of the distribution
[out]	state	state is a complex*16 array of dimension (0:d1−1,0:d2−1) which is a random complex matrix

Implementation

```
subroutine RANDDISTGAUSSC(d1,d2,mu,sigma,state)
implicit none
integer::d1,d2,i,j
complex*16::state(d1,d2)
real*8::rn1,rn2,pi,n1,n2,rn3,rn4,mu,sigma
pi=4.0d0*atan(1.0d0)
do i=1,d1
    do j=1,d2
        call random_number(rn1)
        call random_number(rn2)
        call random_number(rn3)
        call random_number(rn4)
        n1=mu+sigma*sqrt(-2*log(rn1))*cos(2*pi*rn2)
```

```
        n2=mu+sigma*sqrt(-2*log(rn3))*cos(2*pi*rn4)
        state(i,j)=dcmplx(n1,n2)
    end do
end do
end subroutine
```

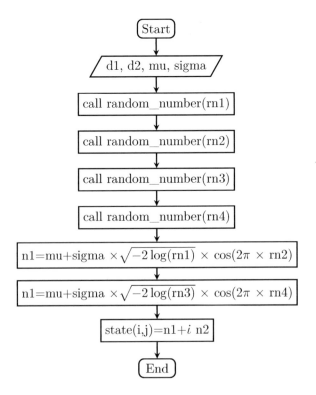

FIGURE 7.2: Flowchart of RANDDISTGAUSSC

Example

```
program example
implicit none
integer,parameter::d=5
integer::i,j
complex*16::state(d,1)
call RANDDISTGAUSSC(d,1,0.0d0,1.d0,state)
end program
```

In the above example, state is a 5×1 complex matrix which stores the random numbers from the Gaussian distribution with zero mean and unit standard deviation.

7.2 Random number generator for uniform distribution in the range (a,b)

The uniform random number generator generates a random number in the arbitrary range (a, b). Let us generate random number r_n between 0 and 1 using the intrinsic fortran function CALL RANDOM_NUMBER(). Now we transform r_n into the range (a, b) as shown below:

$$r_n = a + (b - a)r_n. \tag{7.3}$$

7.2.1 For a real random matrix whose elements are chosen from uniform distribution

---

subroutine RANDDISTUNIR(a,b,d1,d2,state)

Parameters

In/Out	Argument	Description
[in]	a	a (integer) is the lowest value of the range (a,b)
[in]	b	b (integer) is the highest value of the range (a,b)
[in]	d1	d1 (integer) is the number of rows in the final matrix named as state
[in]	d2	d2 (integer) is the number of columns in the final matrix named as state
[out]	state	state is a real*8 array of dimension $(0:d1-1,0:d2-1)$ which is a random real matrix

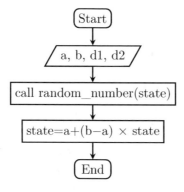

FIGURE 7.3: Flowchart of RANDDISTUNIR

Implementation

```
subroutine RANDDISTUNIR(a,b,d1,d2,state)
implicit none
integer::d1,d2
real*8::state(0:d1-1,0:d2-1),a,b
call random_number(state)
state=a+(b-a)*state
end subroutine
```

Example

```
program example
implicit none
integer,parameter::d=10
integer::i
real*8::state(0:d-1,0:0)
call RANDDISTUNIR(-10.0d0,10.0d0,d,1,state)
end program
```

In the above example, state is a 10×1 real matrix which stores the random numbers chosen from a uniform distribution in range $(-10,10)$.

7.2.2 For a complex random matrix whose elements are chosen from uniform distribution

```
subroutine RANDDISTUNIC(a,b,d1,d2,state)
```

Parameters

In/Out	Argument	Description
[in]	a	a (integer) is the lowest value of the range (a,b)
[in]	b	b (integer) is the highest value of the range (a,b)
[in]	d1	d1 (integer) is the number of rows in the final matrix named as state
[in]	d2	d2 (integer) is the number of columns in the final matrix named as state
[out]	state	state is a complex*16 array of dimension $(0:d1-1,0:d2-1)$ which is a random complex matrix

Implementation

```
subroutine RANDDISTUNIC(a,b,d1,d2,state)
implicit none
```

```
integer::d1,d2,i,j
real*8::state1(0:d1-1,0:d2-1),state2(0:d1-1,0:d2-1),a,b
complex*16::state(0:d1-1,0:d2-1)
call random_number(state1)
call random_number(state2)
state1=a+(b-a)*state1
state2=a+(b-a)*state2
do i=0,d1-1
    do j=0,d2-1
        state(i,j)=dcmplx(state1(i,j),state2(i,j))
    end do
end do
end subroutine
```

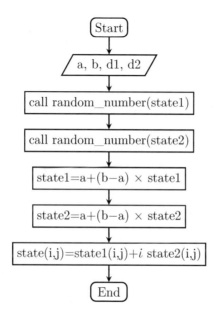

FIGURE 7.4: Flowchart of RANDDISTUNIC

Example

```
program example
implicit none
integer,parameter::d=10
integer::i
complex*16::state(d,d)
call RANDDISTUNIC(10.0d0,200.0d0,d,d,state)
end program
```

In the above example, state is a 10×10 complex matrix which stores the random numbers chosen from a uniform distribution in the range $(10, 200)$.

7.3   Random real symmetric matrices

Any random real symmetric matrix B can be generated from any general matrix A whose elements are chosen from a Gaussian or a uniform distribution as follows,

$$B = \frac{1}{2}(A + A^T) \qquad (7.4)$$

---

subroutine RANDSR(d,dist,a,b,mu,sigma,state)

Parameters

In/Out	Argument	Description
[in]	d	d (integer) is the dimension of the matrix named as state
[in]	dist	dist (character*3) is a string defining the type of distribution if dist = 'GAU', Gaussian distribution will be chosen; if dist = 'UNI', Uniform distribution will be chosen
[in]	a	If dist = 'UNI', a (integer) is the lowest value of the range (a,b); If dist = 'GAU', a is not referenced
[in]	b	If dist = 'UNI', b (integer) is the highest value of the range (a,b); If dist = 'GAU', b is not referenced
[in]	mu	If dist = 'GAU', mu (real*8) is the mean of the distribution; If dist = 'UNI', mu is not referenced
[in]	sigma	If dist = 'GAU', sigma (real*8) is the standard deviation of the distribution; If dist = 'UNI', sigma is not referenced
[out]	state	state is a real*8 array of dimension (0:d−1,0:d−1) which is a random real symmetric matrix

Implementation

```
subroutine RANDSR(d,dist,a,b,mu,sigma,state)
implicit none
integer::i,d
real*8::state(0:d-1,0:d-1),a,b,mu,sigma
character*3::dist
if (dist=='GAU') then
    call RANDDISTGAUSSR(d,d,mu,sigma,state)
end if
if (dist=='UNI') then
```

```
    call RANDDISTUNIR(a,b,d,d,state)
end if
state=(state+transpose(state))/2.d0
end subroutine
```

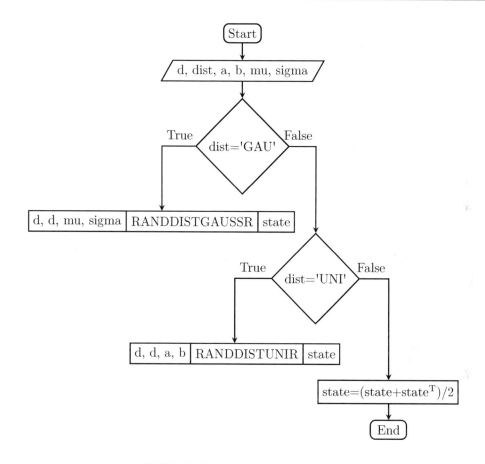

FIGURE 7.5: Flowchart of RANDSR

Example

```
program example
implicit none
integer,parameter::d=3
integer::i,j
real*8::state(d,d),a,b,mu,sigma
character*3::dist
dist='GAU'
mu=0.0d0
sigma=1.0d0
call RANDSR(d,dist,a,b,mu,sigma,state)
end program
```

In the above example, state is a 3×3 real symmetric matrix which stores the random numbers selectively chosen from the Gaussian distribution with mean zero and unit standard deviation.

7.4   Random complex Hermitian matrices

Any random complex Hermitian matrix B can be generated from any complex general matrix A whose elements are chosen from a Gaussian or a uniform distribution as follows,

$$B = \frac{1}{2}(A + A^{\dagger}) \tag{7.5}$$

```
subroutine RANDHC(d,dist,a,b,mu,sigma,state)
```

Parameters

In/Out	Argument	Description
[in]	d	d (integer) is the dimension of the matrix named as state
[in]	dist	dist (character*3) is a string defining the type of distribution if dist = 'GAU', Gaussian distribution will be chosen; if dist = 'UNI', Uniform distribution will be chosen
[in]	a	If dist = 'UNI', a (integer) is the lowest value of the range (a,b); If dist = 'GAU', a is not referenced
[in]	b	If dist = 'UNI', b (integer) is the highest value of the range (a,b); If dist = 'GAU', b is not referenced
[in]	mu	If dist = 'GAU', mu (real*8) is the mean of the distribution; If dist = 'UNI', mu is not referenced
[in]	sigma	If dist = 'GAU', sigma (real*8) is the standard deviation of the distribution; If dist = 'UNI', sigma is not referenced
[out]	state	state is a real*8 array of dimension (0:d−1,0:d−1) which is a random complex Hermitian matrix.

Implementation

```
subroutine RANDHC(d,dist,a,b,mu,sigma,state)
implicit none
integer::i,d
complex*16::state(0:d-1,0:d-1)
real*8::a,b,mu,sigma
```

```
character*3::dist
if (dist=='GAU') then
    call RANDDISTGAUSSC(d,d,mu,sigma,state)
end if
if (dist=='UNI') then
    call RANDDISTUNIC(a,b,d,d,state)
end if
state=(state+transpose(conjg(state)))/2.d0
end subroutine
```

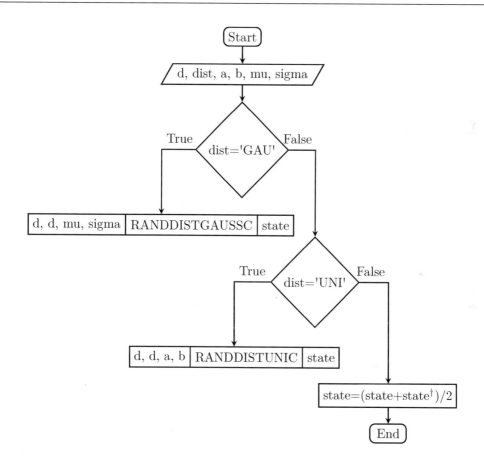

FIGURE 7.6: Flowchart of RANDHC

Example

```
program example
implicit none
integer,parameter::d=2
integer::i,j
real*8::a,b,mu,sigma
complex*16::state(d,d)
character*3::dist
dist='UNI'
```

```
a=10.0d0
b=20.0d0
call RANDHC(d,dist,a,b,mu,sigma,state)
end program
```

In the above example, state is a 3×3 complex Hermitian matrix which stores the random numbers selectively chosen from a uniform distribution in range (10, 20).

7.5   Random unitary matrices

For generating a random unitary matrix B, generate any general real random matrix A with elements chosen from a Gaussian or a uniform distribution. Now, Perform a QR decomposition on the matrix A which will give us a random unitary matrix B such that, $BB^{\dagger} = B^{\dagger}B = \mathbb{I}$, in the case of B being a real matrix, then we will be generating a random orthogonal matrix.

7.5.1   Random real orthogonal matrix

subroutine RANDORTHOGONAL(d1,d2,dist,a,b,mu,sigma,state)

Parameters

In/Out	Argument	Description
[in]	d1	d1 (integer) is the number of rows in the final matrix named as state
[in]	d2	d2 (integer) is the number of columns in the final matrix named as state
[in]	dist	dist (character*3) is a string defining the type of distribution if dist = 'GAU', Gaussian distribution will be chosen; if dist = 'UNI', Uniform distribution will be chosen
[in]	a	If dist = 'UNI', a (integer) is the lowest value of the range (a,b); If dist = 'GAU', a is not referenced
[in]	b	If dist = 'UNI', b (integer) is the highest value of the range (a,b); If dist = 'GAU', b is not referenced
[in]	mu	If dist = 'GAU', mu (real*8) is the mean of the distribution; If dist = 'UNI', mu is not referenced

In/Out	Argument	Description
[in]	sigma	If dist = 'GAU', sigma (real*8) is the standard deviation of the distribution; If dist = 'UNI', sigma is not referenced
[out]	state	state is a real*8 array of dimension (0:d1−1,0:d2−1) which is a random real orthogonal matrix

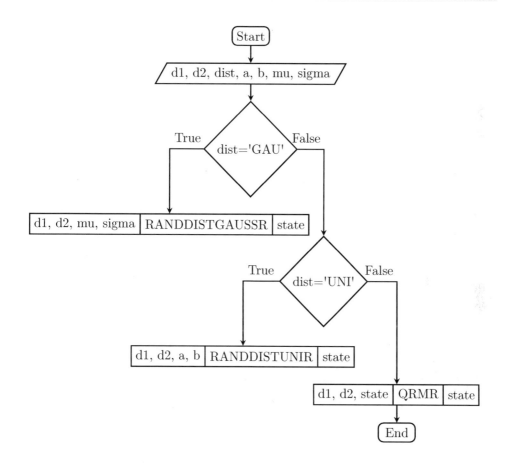

FIGURE 7.7: Flowchart of RANDORTHOGONAL

Implementation

```
subroutine RANDORTHOGONAL(d1,d2,dist,a,b,mu,sigma,state)
implicit none
integer::i,d1,d2
real*8::state(0:d1-1,0:d2-1),a,b,mu,sigma,R(0:d1-1,0:d2-1)
character*3::dist
if (dist=='GAU') then
    call RANDDISTGAUSSR(d1,d2,mu,sigma,state)
end if
```

```
if (dist=='UNI') then
    call RANDDISTUNIR(a,b,d1,d2,state)
end if
call QRMR(d1,d2,state,R)
end subroutine
```

Example

```
program example
implicit none
integer,parameter::d=3
integer::i,j
real*8::state(d,d),a,b,mu,sigma,state2(d,d)
character*3::dist
dist='GAU'
mu=0.0d0
sigma=1.0d0
call RANDORTHOGONAL(d,d,dist,a,b,mu,sigma,state)
end program
```

In the above example, state is a 3×3 real orthogonal matrix which stores the random numbers selectively chosen from the Gaussian distribution with zero mean and unit standard deviation.

7.5.2 Random complex unitary matrices

```
subroutine RANDUNITARY(d1,d2,dist,a,b,mu,sigma,state)
```

Parameters

In/Out	Argument	Description
[in]	d1	d1 (integer) is the number of rows in the final matrix named as state
[in]	d2	d2 (integer) is the number of columns in the final matrix named as state
[in]	dist	dist (character*3) is string defining the type of distribution if dist = 'GAU', Gaussian distribution will be chosen; if dist = 'UNI', Uniform distribution will be chosen
[in]	a	If dist = 'UNI', a (integer) is the lowest value of the range (a,b); If dist = 'GAU', a is not referenced
[in]	b	If dist = 'UNI', b (integer) is the highest value of the range (a,b); If dist = 'GAU', b is not referenced

In/Out	Argument	Description
[in]	mu	If dist = 'GAU', mu (real*8) is the mean of the distribution; If dist = 'UNI', mu is not referenced
[in]	sigma	If dist = 'GAU', sigma (real*8) is the standard deviation of the distribution; If dist = 'UNI', sigma is not referenced
[out]	state	state is a complex array of dimension $(0:d1-1,0:d2-1)$ which is a random complex unitary matrix

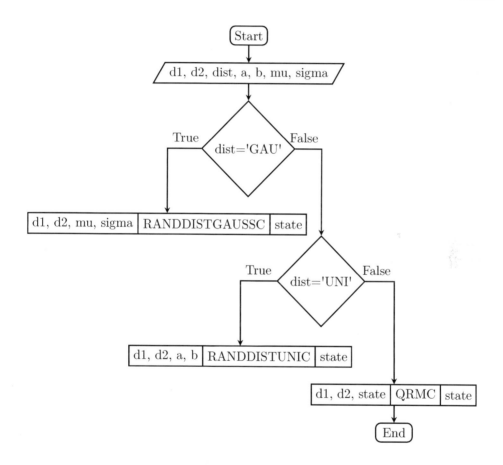

FIGURE 7.8: Flowchart of RANDUNITARY

Implementation

```
subroutine RANDUNITARY(d1,d2,dist,a,b,mu,sigma,state)
implicit none
integer::i,d1,d2
complex*16::state(0:d1-1,0:d2-1),R(0:d1-1,0:d2-1)
real*8::a,b,mu,sigma
```

```
character*3::dist
if (dist=='GAU') then
    call RANDDISTGAUSSC(d1,d2,mu,sigma,state)
end if
if (dist=='UNI') then
    call RANDDISTUNIC(a,b,d1,d2,state)
end if
call QRMC(d1,d2,state,R)
end subroutine
```

Example

```
program example
implicit none
integer,parameter::d=2
integer::i,j
real*8::a,b,mu,sigma
complex*16::state(d,d),state2(d,d)
character*3::dist
dist='UNI'
a=1.0d0
b=2.0d0
call RANDUNITARY(d,d,dist,a,b,mu,sigma,state)
end program
```

In the above example, state is a 2×2 complex unitary matrix which stores the random numbers selectively chosen from a uniform distribution in range $(1, 2)$.

7.6   Real Ginibre matrices

The real Ginibre matrix is a random matrix with the elements chosen from an independent and identically distributed random variables chosen from the normal distribution $\mathcal{N}(0,1)$.

```
subroutine GINIBRER(d,state)
```

Parameters

In/Out	Argument	Description
[in]	d	d (integer) is the dimension of the matrix named as state
[out]	state	state is a real*8 array of dimension $(0:d-1,0:d-1)$ which is a real Ginibre matrix

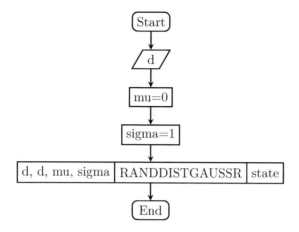

FIGURE 7.9: Flowchart of GINIBRER

Implementation

```
subroutine GINIBRER(d,state)
implicit none
integer::d
real*8::state(0:d-1,0:d-1),mu,sigma
mu=0.0d0
sigma=1.0d0
call RANDDISTGAUSSR(d,d,mu,sigma,state)
end subroutine
```

Example

```
program example
implicit none
integer,parameter::d=3
integer::i,j
real*8::state(d,d)
call GINIBRER(d,state)
end program
```

In the above example, state is a 3×3 required real Ginibre matrix.

7.7 Complex Ginibre matrices

The complex Ginibre matrix is a random matrix with each of the real and imaginary elements individually chosen from an independent and identically distributed random variables chosen from the normal distribution $\mathcal{N}(0, \frac{1}{2})$.

```
subroutine GINIBREC(d,state)
```

Parameters

In/Out	Argument	Description
[in]	d	d (integer) is the dimension of the matrix named as state
[out]	state	state is a complex*16 array of dimension (0:d−1,0:d−1) which is a complex Ginibre matrix

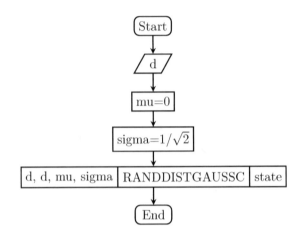

FIGURE 7.10: Flowchart of GINIBREC

Implementation

```
subroutine GINIBREC(d,state)
implicit none
integer::d
complex*16::state(0:d-1,0:d-1)
real*8::mu,sigma
mu=0.0d0
sigma=1.0d0/sqrt(2.0d0)
call RANDDISTGAUSSC(d,d,mu,sigma,state)
end subroutine
```

Example

```
program example
implicit none
integer,parameter::d=3
integer::i,j
complex*16::state(d,d)
call GINIBREC(d,state)
end program
```

In the above example, state is a 3×3 required complex Ginibre matrix.

7.8   Wishart matrices

Wishart matrices are positive definite matrices which can be constructed using the Ginibre matrix G as follows.

$$W = GG^{\dagger} \tag{7.6}$$

7.8.1   Real Wishart matrix

---

subroutine WISHARTSR(d,state)

Parameters

In/Out	Argument	Description
[in]	d	d (integer) is the dimension of the matrix named as state
[out]	state	state is a real*8 array of dimension (0:d−1,0:d−1) which is a real Wishart matrix

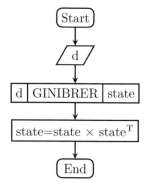

FIGURE 7.11: Flowchart of WISHARTSR

Implementation

```
subroutine WISHARTSR(d,state)
implicit none
integer::d
real*8::state(0:d-1,0:d-1)
call GINIBRER(d,state)
state=matmul(state,transpose(state))
end subroutine
```

Example

```
program example
implicit none
integer,parameter::d=3
integer::i,j
real*8::state(d,d)
call WISHARTSR(d,state)
end program
```

In the above example, state is a 3×3 required real Wishart matrix.

7.8.2 Complex Wishart matrix

```
subroutine WISHARTHC(d,state)
```

Parameters

In/Out	Argument	Description
[in]	d	d (integer) is the dimension of the matrix named as state
[out]	state	state is a complex*16 array of dimension $(0\!:\!d\!-\!1,0\!:\!d\!-\!1)$ which is a complex Wishart matrix

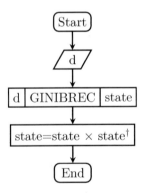

FIGURE 7.12: Flowchart of WISHARTHC

Implementation

```
subroutine WISHARTHC(d,state)
implicit none
integer::d
complex*16::state(0:d-1,0:d-1)
call GINIBREC(d,state)
state=matmul(state,transpose(conjg(state)))
end subroutine
```

Example

```
program example
implicit none
integer,parameter::d=2
integer::i,j
complex*16::state(d,d)
call WISHARTHC(d,state)
end program
```

In the above example, state is a 3×3 required complex Wishart matrix.

7.9   Random probability vector

A random probability vector whose entries are probabilities can be given by,

$$P = (p_1, p_2, \cdots, p_n) \tag{7.7}$$

here, $0 \leq p_i \leq 1$, also the p_i's are chosen from a uniform distribution between the range (0,1). Note that, we have to normalize the vector P such that $\sum_{i=1}^{n} p_i = 1$.

Parameters

```
subroutine RANDPVEC(d,state)
```

In/Out	Argument	Description
[in]	d	d (integer) is the dimension of the one-dimensional array named as state
[out]	state	state is a real*8 array of dimension (0:d−1) which is the column vector containing the probabilities

Implementation

```
subroutine RANDPVEC(d,state)
implicit none
integer::d,i
real*8::state(0:d-1),norm
character*3::dist
call random_number(state)
norm=0.0d0
do i=0,d-1
    norm=norm+state(i)
end do
state=state/norm
end subroutine
```

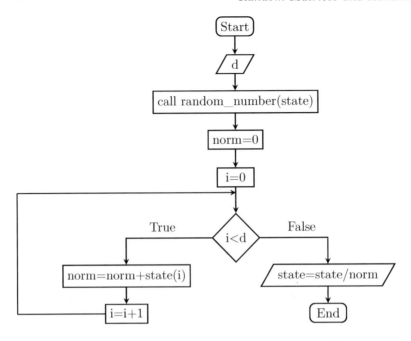

FIGURE 7.13: Flowchart of RANDPVEC

Example

```
program example
implicit none
integer,parameter::d=5
integer::i,j
real*8::state(d),summ
call RANDPVEC(d,state)
end program
```

In the above example, state is a 5×1 required probability vector which is normalized.

7.10   Random pure quantum state vector

The random quantum state vector [115, 122] of dimension n can be described as,

$$|\psi\rangle = \sum_{j=1}^{n} c_j|j\rangle, \tag{7.8}$$

where c_j's are the random numbers, either real or complex chosen from any random distribution, here we use the uniform and Gaussian distribution, such that $\sum_{j=1}^{n} |c_j|^2 = 1$. Note that, $|j\rangle$ is the computational basis. Random quantum states are solely defined analytically based on the normalization condition only which is identical to the equation of a n-dimensional unit sphere. This probability distribution is proportional to the Dirac delta function such that the $P(c_1, c_2, \ldots, c_n) \propto \delta(\sum_{j=1}^{n} |c_j|^2 - 1)$.

7.10.1   Random real state vector

---

subroutine RANDVR(d,dist,a,b,mu,sigma,state)

Parameters

In/Out	Argument	Description
[in]	d	d (integer) is the dimension of the one-dimensional array named as state
[in]	dist	dist (character*3) is string defining the type of distribution if dist = 'GAU', Gaussian distribution will be chosen; if dist = 'UNI', Uniform distribution will be chosen
[in]	a	If dist = 'UNI', a (integer) is the lowest value of the range (a,b); If dist = 'GAU', a is not referenced
[in]	b	If dist = 'UNI', b (integer) is the highest value of the range (a,b); If dist = 'GAU', b is not referenced
[in]	mu	If dist = 'GAU', mu (real*8) is the mean of the distribution; If dist = 'UNI', mu is not referenced
[in]	sigma	If dist = 'GAU', sigma (real*8) is the standard deviation of the distribution; If dist = 'UNI', sigma is not referenced
[out]	state	state is a real*8 array of dimension (0:d−1) which is the column vector giving the random real state vector

Implementation

```
subroutine RANDVR(d,dist,a,b,mu,sigma,state)
implicit none
integer::i,d
real*8::state1(0:d-1,0:0),a,b,state(0:d-1),norm,mu,sigma
character*3::dist
if (dist=='GAU') then
    call RANDDISTGAUSSR(d,1,mu,sigma,state1)
end if
if (dist=='UNI') then
    call RANDDISTUNIR(a,b,d,1,state1)
end if
norm=0.0d0
do i=0,d-1,1
    state(i)=state1(i,0)
end do
call NORMALIZEVR(state,d)
```

```
end subroutine
```

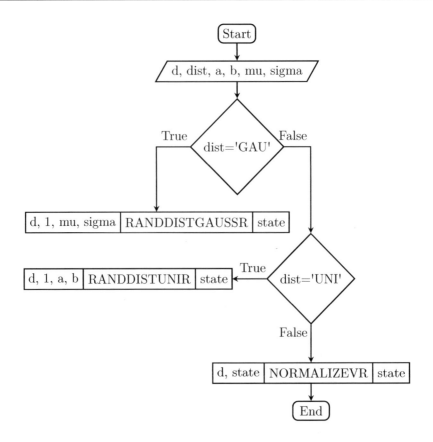

FIGURE 7.14: Flowchart of RANDVR

Example

```
program example
implicit none
integer,parameter::d=5
integer::i,j
real*8::state(d),a,b,mu,sigma,norm
character*3::dist
dist='GAU'
mu=0.0d0
sigma=1.0d0
call RANDVR(d,dist,a,b,mu,sigma,state)
end program
```

In the above example, state is a 5×1 real random quantum state vector whose entries are selectively chosen from the Gaussian distribution with zero mean and unit standard deviation.

7.10.2 Random pure complex state vector

---

subroutine RANDVC(d,dist,a,b,mu,sigma,state)

Parameters

In/Out	Argument	Description
[in]	d	d (integer) is the dimension of the one-dimensional array named as state
[in]	dist	dist (character*3) is string defining the type of distribution if dist = 'GAU', Gaussian distribution will be chosen; if dist = 'UNI', Uniform distribution will be chosen
[in]	a	If dist = 'UNI', a (integer) is the lowest value of the range (a,b); If dist = 'GAU', a is not referenced
[in]	b	If dist = 'UNI', b (integer) is the highest value of the range (a,b); If dist = 'GAU', b is not referenced
[in]	mu	If dist = 'GAU', mu (real*8) is the mean of the distribution; If dist = 'UNI', mu is not referenced
[in]	sigma	If dist = 'GAU', sigma (real*8) is the standard deviation of the distribution; If dist = 'UNI', sigma is not referenced
[out]	state	state is a complex*16 array of dimension (0:d−1) which is the column vector giving the random complex state vector

Implementation

```
subroutine RANDVC(d,dist,a,b,mu,sigma,state)
implicit none
integer::i,d
complex*16::state1(0:d-1,0:0),state(0:d-1)
real*8::a,b,mu,sigma
character*3::dist
if (dist=='GAU') then
    call RANDDISTGAUSSC(d,1,mu,sigma,state1)
end if
if (dist=='UNI') then
    call RANDDISTUNIC(a,b,d,1,state1)
end if
do i=0,d-1,1
    state(i)=state1(i,0)
end do
call NORMALIZEVC(state,d)
```

`end subroutine`

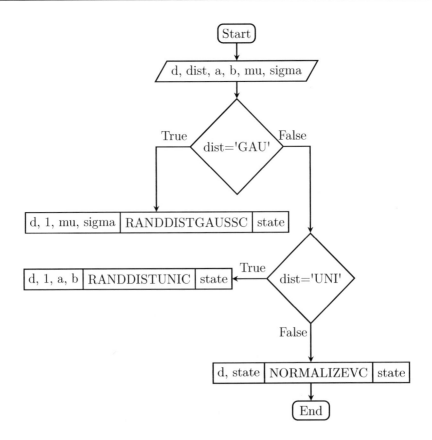

FIGURE 7.15: Flowchart of RANDVC

Example

```
program example
implicit none
integer,parameter::d=5
integer::i,j
real*8::a,b,mu,sigma
complex*16::state(d)
character*3::dist
dist='GAU'
mu=0.0d0
sigma=1.0d0
call RANDVC(d,dist,a,b,mu,sigma,state)
do i=1,d
   write(111,*)state(i)
end do
end program
```

In the above example, state is a 5×1 complex random pure quantum state vector whose elements are selectively chosen from the Gaussian distribution with zero mean and unit standard deviation.

7.11 Random density matrices

A random density matrix [118, 119] ρ can be generated using a real or a complex Ginibre matrix G as follows,

$$\rho = \frac{GG^\dagger}{Tr(GG^\dagger)} \tag{7.9}$$

The matrix ρ is by construction Hermitian, positive definite and normalized. The ensemble so generated is called the Hilbert–Schmidt ensemble.

7.11.1 Random real density matrix

subroutine RANDDMR(d,state)

Parameters

In/Out	Argument	Description
[in]	d	d (integer) is the dimension of the matrix named as state
[out]	state	state is a real*8 array of dimension (0:d−1,0:d−1) which is the random real density matrix

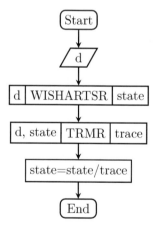

FIGURE 7.16: Flowchart of RANDDMR

Implementation

```
subroutine RANDDMR(d,state)
integer::d
real*8::state(0:d-1,0:d-1),mu,sigma,trace
call WISHARTSR(d,state)
trace=0.0d0
do i=0,d-1,1
    trace=trace+state(i,i)
end do
state=state/trace
end subroutine
```

Example

```
program example
implicit none
integer,parameter::d=3
real*8::state(d,d)
call RANDDMR(d,state)
end program
```

In the above example, state is a required 3×3 real density matrix.

7.11.2   Random complex density matrix

```
subroutine RANDDMC(d,state)
```

Parameters

In/Out	Argument	Description
[in]	d	d (integer) is the dimension of the matrix named as state
[out]	state	state is a complex*16 array of dimension $(0:d-1,0:d-1)$ which is the random complex density matrix

Implementation

```
subroutine RANDDMC(d,state)
integer::d
complex*16::state(0:d-1,0:d-1),trace
call WISHARTHC(d,state)
trace=0.0d0
do i=0,d-1,1
trace=trace+state(i,i)
end do
state=state/trace
end subroutine
```

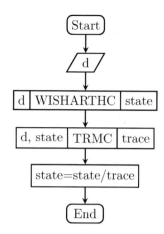

FIGURE 7.17: Flowchart of RANDDMC

Example

```
program example
implicit none
integer,parameter::d=3
complex*16::state(d,d)
call RANDDMC(d,state)
end program
```

In the above example, state is a required 3×3 complex density matrix.

Dependency Chart

In this part of the book, we will give the dependency chart for all the recipes contained in the book. It is given as a branching diagram wherein the subroutine or the recipe under consideration occurs to the left most part of the figure and the figure branches in the left to right direction carrying the dependencies of the main routine and the other routines which are an integral part of the main routine. This diagrammatic representation becomes important for the user to decide what recipes he has to choose so that the main routine will run without any glitches.

Chapter 2

IPVR

$\boxed{\text{IPVR}}$

IPVC

$\boxed{\text{IPVC}}$

NORMVR

$\boxed{\text{NORMVR}}$

NORMVC

$\boxed{\text{NORMVC}}$

NORMALIZEVR

$\boxed{\text{NORMALIZEVR}}$

NORMALIZEVC

$\boxed{\text{NORMALIZEVC}}$

OPVR

$\boxed{\text{OPVR}}$

OPVC

$\boxed{\text{OPVC}}$

MMULMR

$\boxed{\text{MMULMR}}$

MMULMC

$\boxed{\text{MMULMC}}$

KPMR

$\boxed{\text{KPMR}}$

KPMC

$\boxed{\text{KPMC}}$

TRMR

$\boxed{\text{TRMR}}$

TRMC

$\boxed{\text{TRMC}}$

COMMR

$\boxed{\text{COMMR}}$

COMMC

$\boxed{\text{COMMC}}$

ANTICOMMR

$\boxed{\text{ANTICOMMR}}$

ANTICOMMC

$\boxed{\text{ANTICOMMC}}$

BTOD

$\boxed{\text{BTOD}}$

DTOB

DTOB

DTOBONEBIT

DTOBONEBIT——DTOB

LEFTSHIFT

LEFTSHIFT

RIGHTSHIFT

RIGHTSHIFT

SWAP

SWAP

Chapter 3

MATINVMR

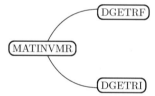

MATINVMC

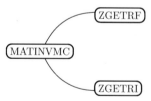

LANCZOSSR

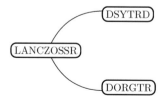

LANCZOSHC

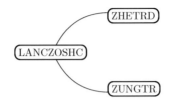

QRMR

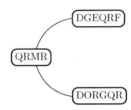

QRMC

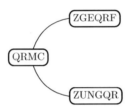

FUNSR

FUNHC

FUNMR

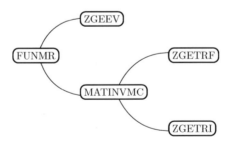

FUNMC

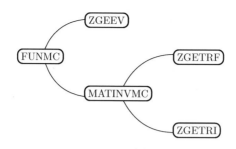

POWNSR

POWNSR——DSYEV

POWNHC

POWNHC——ZHEEV

POWNMR

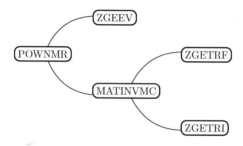

POWNMC

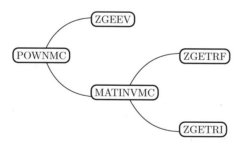

TRPOWNSR

TRPOWNSR——DSYEV

TRPOWNHC

TRPOWNHC——ZHEEV

TRPOWNMR

TRPOWNMR —— DGEEV

TRPOWNMC

TRPOWNMC —— ZGEEV

DETSR

DETSR —— DSYEV

DETHC

DETHC —— ZHEEV

DETMR

DETMR —— DGEEV

DETMC

DETMC —— ZGEEV

TRNORMMR

TRNORMMR —— TRPOWNSR —— DSYEV

TRNORMMC

TRNORMMC —— TRPOWNHC —— ZHEEV

HISCNORMMR

HISCNORMMR —— TRMR

HISCNORMMC

HISCNORMMC —— TRMC

ABSMR

ABSMR —— POWNSR —— DSYEV

ABSMC

ABSMC —— POWNHC —— ZHEEV

HISCIPMR

HISCIPMR——TRMR

HISCIPMC

HISCIPMC——TRMC

GSOMR

IPVR

GSOMR

NORMALIZEVR

GSOMC

IPVC

GSOMC

NORMALIZEVC

SVDMR

SVDMR——DGESVD

SVDMC

SVDMC——ZGESVD

Chapter 4

PAULIX

PAULIX

PAULIY

PAULIY

PAULIZ

PAULIZ

HADAMARDG

$$\boxed{\text{HADAMARDG}}$$

PHASEG

$$\boxed{\text{PHASEG}}$$

ROTG

$$\boxed{\text{ROTG}}$$

CONOTG

$$\boxed{\text{CONOTG}}$$

COZG

$$\boxed{\text{COZG}}$$

SWAPG

$$\boxed{\text{SWAPG}}$$

TOFFOLIG

$$\boxed{\text{TOFFOLIG}}$$

FREDKING

$$\boxed{\text{FREDKING}}$$

NQBHADAMARDG

$$\boxed{\text{NQBHADAMARDG}}\!-\!\boxed{\text{DTOB}}$$

NQBCB

$$\boxed{\text{NQBCB}}$$

BELL

$$\boxed{\text{BELL}}$$

NQBGHZ

$$\boxed{\text{NQBGHZ}}$$

NQBWSTATE

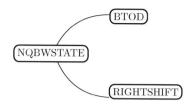

NQBWERNER

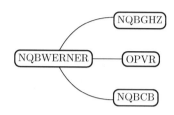

SE

LEDMR

LEDMR————TRPOWNSR————DSYEV

LEDMC

LEDMC————TRPOWNHC————ZHEEV

REDMR

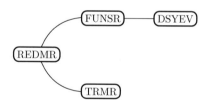

REDMC

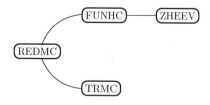

TRDISDMR

TRDISDMR — TRNORMMR — TRPOWNSR — DSYEV

TRDISDMC

TRDISDMC — TRNORMMC — TRPOWNHC — ZHEEV

FIDVR

FIDVR — IPVR

FIDVC

FIDVC — IPVC

FIDDMR

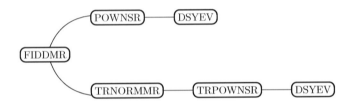

FIDDMR
 ├─ POWNSR — DSYEV
 └─ TRNORMMR — TRPOWNSR — DSYEV

FIDDMC

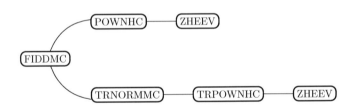

FIDDMC
 ├─ POWNHC — ZHEEV
 └─ TRNORMMC — TRPOWNHC — ZHEEV

FIDVRDMR

FIDVRDMR — IPVR

FIDVCDMC

FIDVCDMC — IPVC

SUPFIDDMR

SUPFIDDMR — TRMR

SUPFIDDMC

SUPFIDDMC — TRMC

BURDISVR

BURDISVC

BURDISDMR

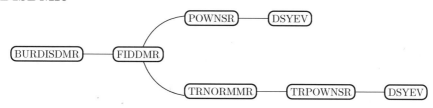

BURDISDMC

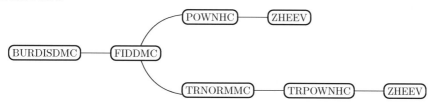

AVGVRMR

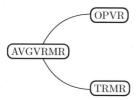

AVGVCMC

AVGDMRMR

AVGDMCMC

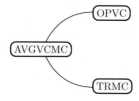

MEASUREVR

MEASUREVR——IPVR

MEASUREVC

MEASUREVC——IPVC

MEASUREDMR

MEASUREDMR——TRMR

MEASUREDMC

MEASUREDMC——TRMC

Chapter 5
PTRVR

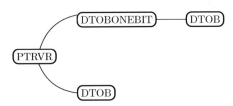

PTRVC

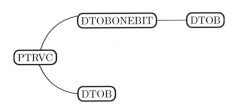

PTRDMR

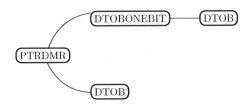

PTRDMC

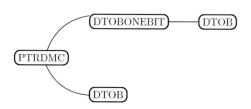

PTDMR

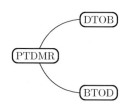

PTDMC

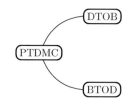

PTVR

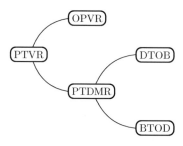

PTVC

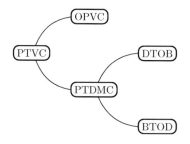

CONVR

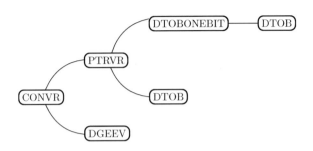

CONVC

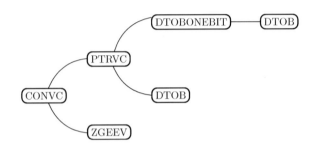

CONDMR

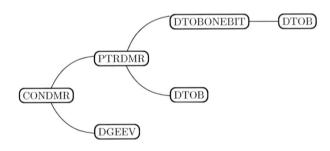

CONDMC

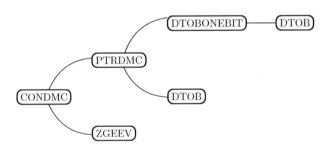

BERPVR

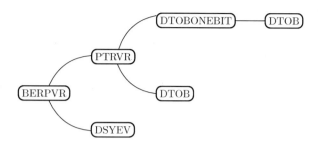

BERPVC

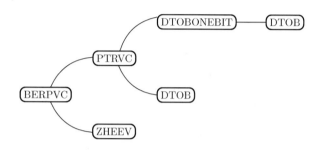

BERPDMR

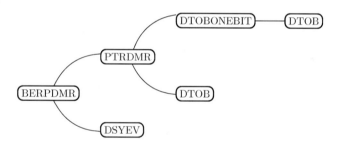

BERPDMC

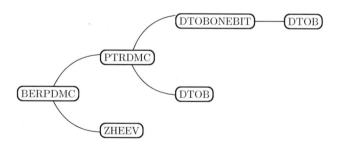

RERPDMR

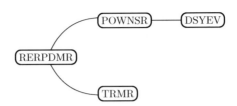

RERPDMC

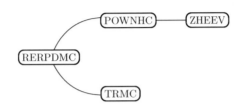

NEGVR

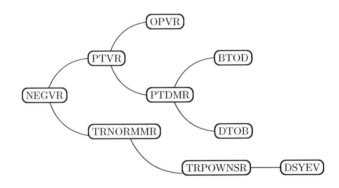

NEGVC

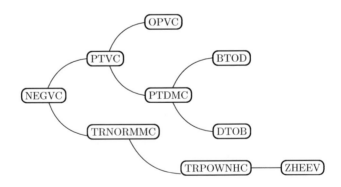

NEGDMR

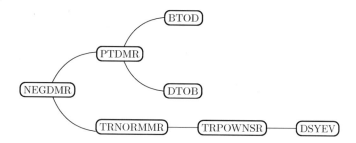

NEGDMC

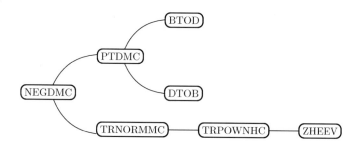

QMVR

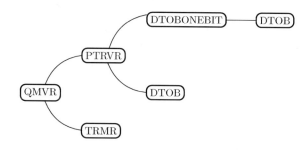

QMVC

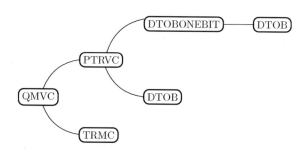

QMDMR

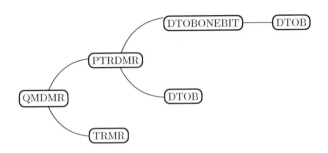

QMDMC

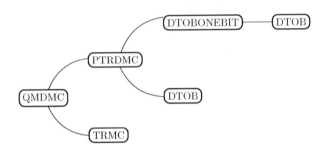

ENTSPECDMR

ENTSPECDMR —— DSYEV

ENTSPECDMC

ENTSPECDMC —— ZHEEV

RESENTVR

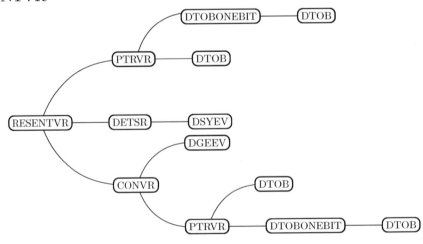

RESENTVC

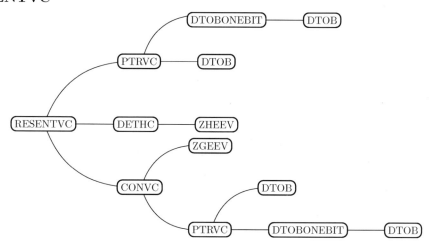

RESENTDMR

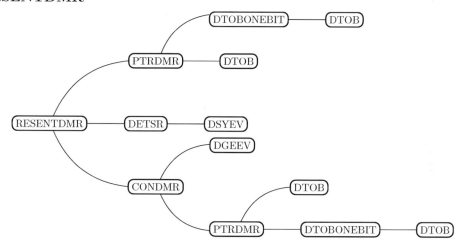

RESENTDMC

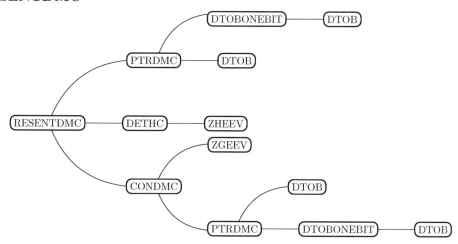

Chapter 6

XHAM

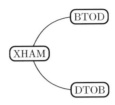

YHAM

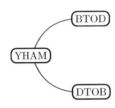

ZHAM

ISINGXHAM

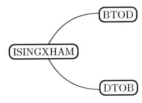

ISINGYHAM

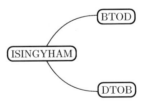

ISINGZHAM

HEIHAM

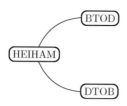

DMXHAM

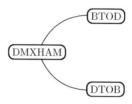

DMYHAM

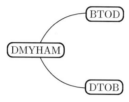

DMZHAM

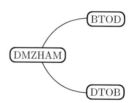

Chapter 7
RANDDISTGAUSSR

RANDDISTGAUSSR

RANDDISTGAUSSC

RANDDISTGAUSSC

RANDDISTUNIR

RANDDISTUNIR

RANDDISTUNIC

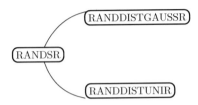

RANDSR

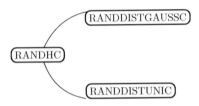

RANDHC

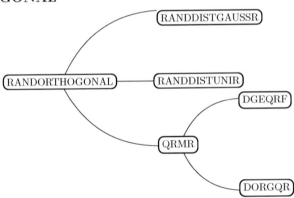

RANDORTHOGONAL

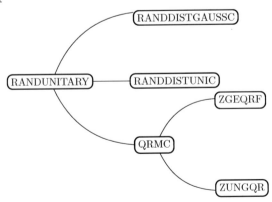

RANDUNITARY

GINIBRER

GINIBREC

WISHARTSR

WISHARTSR —— GINIBRER —— RANDDISTGAUSSR

WISHARTHC

WISHARTHC —— GINIBREC —— RANDDISTGAUSSC

RANDPVEC

RANDPVEC

RANDVR

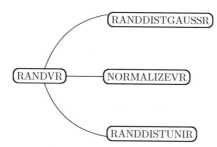

RANDVC

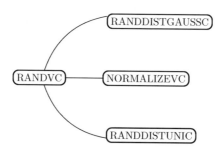

RANDDMR

RANDDMC

Bibliography

[1] V. Rajaraman, *Computer programming in FORTRAN 90 and 95* (PHI Learning Pvt. Ltd. 1997).

[2] S. J. Chapman, *FORTRAN 90/95 for scientists and engineers* (McGraw Hill 1997).

[3] B. Hahn, *FORTRAN 90 for scientists and engineers* (Butterworth-Heinemann 1994).

[4] J. Backus, R. J. Beeber, S. Best, R. Goldberg, L. M. Haibt, H. Herrick, R. Nelson, D. Sayre, P. Sheridan, H. Stern, I. Ziller, R. A. Hughes, and R. Nutt, The FORTRAN automatic coding system (Association for Computing Machinery 1957).

[5] E. Anderson, Z. Bai, C. Bischof, L. S. Blackford, J. Demmel, J. Dongarra, J. Du Croz, A. Greenbaum, S. Hammarling, A. McKenney *et al.*, *LAPACK users' guide* (SIAM 1999).

[6] R. Shankar, *Principles of quantum mechanics* (Springer 2012).

[7] L. I. Schiff, *Quantum mechanics* (McGraw Hill 1969).

[8] N. Zettili, *Quantum mechanics: concepts and applications* (Wiley 2009).

[9] T. F. Jordan, *Quantum mechanics in simple matrix form* (Dover Publications 2005).

[10] J. Baggott, *The Quantum cookbook: mathematical recipes for the foundations of quantum mechanics* (Oxford University Press 2020).

[11] M. A. Nielsen and I. L. Chuang, *Quantum computation and quantum information* (Cambridge University Press 2010).

[12] R. A. Horn and C. R. Johnson, *Matrix analysis* (Cambridge University Press 2012).

[13] G. Strang, *Linear algebra and its applications.* (Cengage Learning 2005).

[14] P. A. M. Dirac, *Lectures on quantum mechanics* (Dover Publications 2003).

[15] J. M. Ziman, *Elements of advanced quantum theory* (Cambridge University Press 1975).

[16] H. Wendland, *Numerical linear algebra: an introduction* (Cambridge University Press 2017).

[17] S. H. Friedberg, A. J. Insel, and L. E. Spence, *Linear algebra* (Pearson 2002).

[18] X.-D. Zhang, *Matrix analysis and applications* (Cambridge University Press 2017).

[19] C. Lanczos, An iteration method for the solution of the eigenvalue problem of linear differential and integral operators, Journal of Research of the National Bureau of Standards 45 (1950).

[20] L. N. Trefethen and D. Bau, *Numerical linear algebra* (SIAM 1997).

[21] N. J. Higham, *Functions of matrices: theory and computation* (SIAM 2008).

[22] M. Malek-Shahmirzadi, A characterization of certain classes of matrix norms, Linear and Multilinear Algebra 13 (1983).

[23] J. H. M. Wedderburn, The absolute value of the product of two matrices, Bulletin of the AMS 31 (1925).

[24] J. B. Conway, *A course in functional analysis* (Springer 2007).

[25] W. Cheney and D. Kincaid, *Linear algebra: theory and applications* (Jones and Bartlett 2011).

[26] L. Pursell and S. Trimble, Gram-Schmidt orthogonalization by Gauss elimination, The American Mathematical Monthly 98 (1991).

[27] K. Cahill, *Physical mathematics* (Cambridge University Press 2013).

[28] N. S. Yanofsky and M. A. Mannucci, *Quantum computing for computer scientists* (Cambridge University Press 2008).

[29] D. P. DiVincenzo, Quantum gates and circuits, Proceedings of the Royal Society of London A 454 (1998).

[30] R. P. Feynman, Quantum mechanical computers, Foundations of Physics 16 (1986).

[31] D. E. Deutsch, Quantum computational networks, Proceedings of the Royal Society of London A 425 (1989).

[32] T. Toffoli, Computation and construction universality of reversible cellular automata, Journal of Computer and System Sciences 15 (1977).

[33] E. Fredkin and T. Toffoli, Conservative logic, International Journal of Theoretical Physics 21 (1982).

[34] D. Deutsch and R. Jozsa, Rapid solution of problems by quantum computation, Proceedings of the Royal Society of London A 439 (1992).

[35] D. Coppersmith, An approximate Fourier transform useful in quantum factoring, arXiv preprint quant-ph/0201067 (2002).

[36] B. Schumacher, Quantum coding, Physical Review A 51 (1995).

[37] S. Aaronson, *Quantum computing since democritus* (Cambridge University Press 2013).

[38] V. Vedral, *Introduction to quantum information science* (Oxford University Press 2007).

[39] C. H. Bennett, G. Brassard, C. Crépeau, R. Jozsa, A. Peres, and W. K. Wootters, Teleporting an unknown quantum state via dual classical and Einstein-Podolsky-Rosen channels, Physical Review Letters 70 (1993).

[40] C. H. Bennett and S. J. Wiesner, Communication via one and two particle operators on Einstein-Podolsky-Rosen states, Physical Review Letters 69 (1992).

[41] D. Bohm and Y. Aharonov, Discussion of experimental proof for the paradox of Einstein, Rosen, and Podolsky, Physical Review 108 (1957).

[42] D. M. Greenberger, M. A. Horne, and A. Zeilinger, *Bell's theorem, quantum theory, and conceptions of the universe* (Springer 1989).

[43] S. Oh, Generation of entanglement in finite spin systems via adiabatic quantum computation, The European Physical Journal D 58 (2010).

[44] W. Dür, G. Vidal, and J. I. Cirac, Three qubits can be entangled in two inequivalent ways, Physical Review A 62 (2000).

[45] M. S. Ramkarthik, D. Tiwari, and P. Barkataki, Quantum discord and logarithmic negativity in the generalized N-qubit Werner state, International Journal of Theoretical Physics 59 (2020).

[46] C. E. Shannon, A mathematical theory of communication, The Bell System Technical Journal 27 (1948).

[47] C. Thomas and T. Joy, *Elements of information theory* (Wiley–Blackwell 2006).

[48] V. Vedral, The role of relative entropy in quantum information theory, Reviews of Modern Physics 74 (2002).

[49] C. Fuchs and J. van de Graaf, Cryptographic distinguishability measures for quantum-mechanical states, Information Theory, IEEE Transactions on Information Theory 45 (1999).

[50] R. Jozsa, Fidelity for mixed quantum states, Journal of Modern Optics 41 (1994).

[51] C. A. Fuchs and C. M. Caves, Ensemble-dependent bounds for accessible information in quantum mechanics, Physical Review Letters 73 (1994).

[52] Z.-H. Chen, Z. Ma, F.-L. Zhang, and J.-L. Chen, Super fidelity and related metrics, Central European Journal of Physics 9, 1036 (2011).

[53] W. K. Wootters, Statistical distance and Hilbert space, Physical Review D 23 (1981).

[54] A. Uhlmann, The "transition probability" in the state space of a *-algebra, Reports on Mathematical Physics 9 (1976).

[55] D. Bures, An extension of Kakutani's theorem on infinite product measures to the tensor product of semifinite w^*-algebras, Transactions of the American Mathematical Society 135 (1969).

[56] C. J. Isham, *Lectures on quantum theory mathematical and structural foundations* (Imperial College Press 1995).

[57] C. Cohen-Tannoudji, B. Diu, and F. Laloe, *Quantum mechanics, vol-1* (Wiley VCH 2006).

[58] V. B. Braginsky and F. Y. Khalili, *Quantum measurement* (Cambridge University Press 1995).

[59] A. S. Holevo, *Statistical structure of quantum theory* (Springer 2001).

[60] R. Horodecki, P. Horodecki, M. Horodecki, and K. Horodecki, Quantum entanglement, Reviews of Modern Physics 81 (2009).

[61] E. Schrödinger, Discussion of probability relations between separated systems, Mathematical Proceedings of the Cambridge Philosophical Society 31 (1935).

[62] E. Schrödinger, Die gegenwärtige situation in der quantenmechanik, Naturwissenschaften 23 (1935).

[63] J. A. Wheeler and W. H. Zurek, *Quantum theory and measurement* (Princeton University Press 2014).

[64] B. M. Terhal, M. M. Wolf, and A. C. Doherty, Quantum entanglement: A modern perspective, Physics Today 56 (2003).

[65] A. Einstein, B. Podolsky, and N. Rosen, Can quantum-mechanical description of physical reality be considered complete?, Physical Review 47 (1935).

[66] S. Sachdev, *Quantum phase transitions* (Cambridge University Press 2011).

[67] A. Osterloh, L. Amico, G. Falci, and R. Fazio, Scaling of entanglement close to a quantum phase transition, Nature 416 (2002).

[68] M. B. Plenio and S. S. Virmani, An introduction to entanglement measures, Quantum Information & Computation 7 (2007).

[69] A. Peres, Separability criterion for density matrices, Physical Review Letters 77 (1996).

[70] L. P. Hughston, R. Jozsa, and W. K. Wootters, A complete classification of quantum ensembles having a given density matrix, Physics Letters A 183 (1993).

[71] O. Gühne and G. Tóth, Entanglement detection, Physics Reports 474 (2009).

[72] P. Barkataki and M. S. Ramkarthik, A set theoretical approach for the partial tracing operation in quantum mechanics, International Journal of Quantum Information 16 (2018).

[73] D. Bruß and C. Macchiavello, How the first partial transpose was written?, Foundations of Physics 35 (2005).

[74] W. K. Wootters, Entanglement of formation of an arbitrary state of two qubits, Physical Review Letters 80 (1998).

[75] W. K. Wootters, Entanglement of formation and concurrence, Quantum Information & Computation 1 (2001).

[76] A. Buchleitner, A. Carvalho, and F. Mintert, Measures and dynamics of entanglement, The European Physical Journal Special Topics 159 (2008).

[77] P. Rungta and C. M. Caves, Concurrence-based entanglement measures for isotropic states, Physical Review A 67 (2003).

[78] A. Wehrl, General properties of entropy, Reviews of Modern Physics 50 (1978).

[79] M. Fagotti and P. Calabrese, Entanglement entropy of two disjoint blocks in XY chains, Journal of Statistical Mechanics: Theory and Experiment 2010 (2010).

[80] A. Rényi, On measures of entropy and information, Proceedings of the Fourth Berkeley Symposium on Mathematical Statistics and Probability (1960).

[81] F. Franchini, A. Its, and V. Korepin, Renyi entropy of the XY spin chain, Journal of Physics A: Mathematical and Theoretical 41 (2007).

[82] G. Vidal and R. F. Werner, Computable measure of entanglement, Physical Review A 65 (2002).

[83] M. B. Plenio, Logarithmic negativity: a full entanglement monotone that is not convex, Physical Review Letters 95 (2005).

[84] G. K. Brennen, An observable measure of entanglement for pure states of multi-qubit systems, Quantum Information & Computation 3 (2003).

[85] D. A. Meyer and N. R. Wallach, Global entanglement in multiparticle systems, Journal of Mathematical Physics 43 (2002).

[86] H. Li and F. D. M. Haldane, Entanglement spectrum as a generalization of entanglement entropy: Identification of topological order in non-abelian fractional quantum hall effect states, Physical Review Letters 101 (2008).

[87] M. S. Ramkarthik, V. R. Chandra, and A. Lakshminarayan, Entanglement signatures for the dimerization transition in the Majumdar-Ghosh model, Physical Review A 87 (2013).

[88] V. Coffman, J. Kundu, and W. K. Wootters, Distributed entanglement, Physical Review A 61 (2000).

[89] M. Horodecki, P. Horodecki, and R. Horodecki, Separability of mixed states: necessary and sufficient conditions, Physics Letters A 223 (1996).

[90] M. Horodecki and P. Horodecki, Reduction criterion of separability and limits for a class of distillation protocols, Physical Review A 59 (1999).

[91] D. C. Mattis, *The many-body problem: an encyclopedia of exactly solved models in one dimension* (World Scientific 1993).

[92] K. Joel, D. Kollmar, and L. F. Santos, An introduction to the spectrum, symmetries, and dynamics of spin-1/2 Heisenberg chains, American Journal of Physics 81 (2013).

[93] J. B. Parkinson and D. J. Farnell, *An introduction to quantum spin systems* (Springer 2010).

[94] R. I. Nepomechi, A spin chain primer, International Journal of Modern Physics B 13 (1999).

[95] A. Auerbach, *Interacting electrons and quantum magnetism* (Springer 1998).

[96] P. Fazekas, *Lecture notes on electron correlation and magnetism* (World scientific 1999).

[97] D. C. Mattis, *The theory of magnetism made simple: an introduction to physical concepts and to some useful mathematical methods* (World Scientific 2006).

[98] T. J. G. Apollaro, G. M. A. Almeida, S. Lorenzo, A. Ferraro, and S. Paganelli, Spin chains for two-qubit teleportation, Physical Review A 100 (2019).

[99] J. P. Barjaktarevic, R. H. McKenzie, J. Links, and G. J. Milburn, Measurement-based teleportation along quantum spin chains, Physical Review Letters 95 (2005).

[100] A. Bayat and S. Bose, Entanglement transfer through an antiferromagnetic spin chain, Advances in Mathematical Physics 2010 (2010).

[101] P. Štelmachovič and V. Bužek, Quantum-information approach to the Ising model: Entanglement in chains of qubits, Physical Review A 70 (2004).

[102] M. Plischke and B. Bergersen, *Equilibrium statistical physics* (World Scientific 2006).

[103] S. G. Brush, History of the Lenz-Ising model, Reviews of Modern Physics 39 (1967).

[104] P. Pfeuty, The one-dimensional Ising model with a transverse field, Annals of Physics 57 (1970).

[105] H. Bethe, Zur theorie der metalle, Zeitschrift für Physik 71 (1931).

[106] S. Katsura, Statistical mechanics of the anisotropic linear Heisenberg model, Physical Review 127 (1962).

[107] W. Heisenberg, Zur theorie des ferromagnetismus, Zeitschrift für Physik 49 (1928).

[108] C. K. Majumdar, Antiferromagnetic model with known ground state, Journal of Physics C: Solid State Physics 3 (1970).

[109] C. K. Majumdar and D. K. Ghosh, On next-nearest-neighbor interaction in linear chain-I, Journal of Mathematical Physics 10 (1969).

[110] C. K. Majumdar and D. K. Ghosh, On next-nearest-neighbor interaction in linear chain- II, Journal of Mathematical Physics 10 (1969).

[111] P. W. Anderson, Antiferromagnetism. theory of superexchange interaction, Physical Review 79 (1950).

[112] T. Moriya, Anisotropic superexchange interaction and weak ferromagnetism, Physical Review 120 (1960).

[113] I. Dzyaloshinsky, A thermodynamic theory of "weak" ferromagnetism of antiferro-magnetics, Journal of Physics and Chemistry of Solids 4 (1958).

[114] F. Schütz, P. Kopietz, and M. Kollar, What are spin currents in Heisenberg magnets?, The European Physical Journal B-Condensed Matter and Complex Systems 41 (2004).

[115] W. K. Wootters, Random quantum states, Foundations of Physics 20 (1990).

[116] P. Forrester, *Log-gases and random matrices* (Princeton University Press 2005).

[117] M. L. Mehta, *Random matrices* (Academic Press 2004).

[118] J. Maziero, Fortran code for generating random probability vectors, unitaries, and quantum states, Frontiers in ICT 3 (2016).

[119] K. Życzkowski, K. A. Penson, I. Nechita, and B. Collins, Generating random density matrices, Journal of Mathematical Physics 52 (2011).

[120] B. Collins and I. Nechita, Random matrix techniques in quantum information theory, Journal of Mathematical Physics 57 (2016).

[121] G. E. Box and M. Muller, A note on the generation of random normal deviates, Annals of Mathematical Statistics 29 (1958).

[122] T. A. Brody, J. Flores, J. B. French, P. A. Mello, A. Pandey, and S. S. M. Wong, *Random-matrix physics: spectrum and strength fluctuations*, Reviews of Modern Physics 53 (1981).

Subject Index

Printed in the United States
by Baker & Taylor Publisher Services